SpringerBriefs in Philosophy

For further volumes:
http://www.springer.com/series/10082

William Eckhardt

Paradoxes in Probability Theory

 Springer

William Eckhardt
Eckhardt Trading Company
Chicago, IL
USA

ISSN 2211-4548   ISSN 2211-4556   (electronic)
ISBN 978-94-007-5139-2   ISBN 978-94-007-5140-8   (eBook)
DOI 10.1007/978-94-007-5140-8
Springer Dordrecht Heidelberg New York London

Library of Congress Control Number: 2012944972

Printed on acid-free paper

Springer is part of Springer Science+Business Media (www.springer.com)

Gentle reader, I have sought to make this the finest, most balanced, most beautiful book possible, but could not subvert the law that like begets like.

Miguel de Cervantes, *Don Quixote*

Acknowledgments

For Useful Suggestions, I would like to thank Dale Dellutri, Robert Fefferman, Frank Hodal, Don Howard, David McAllester, Nicholas Polson, Stephen Stigler, Shmuel Weinberger, and Robert Wald, and for typing and editing, Jody Stickann.

Contents

1   **Seven Paradoxes** 1
    1.1   The Doomsday Argument 1
    1.2   The Betting Crowd 2
    1.3   The Simulation Argument 2
    1.4   Newcomb's Problem 2
    1.5   The Open Box Game 3
    1.6   The Hadron Collider Card Experiment 3
    1.7   The Two-Envelopes Paradox 3
    References 4

Part I   Anthropic Fallacies

2   **DOOMSDAY!** 7
    2.1   Randomness in Reference Class 8
    2.2   Retrocausality 9
    References 10

3   **The Betting Crowd** 11
    3.1   Randomness and Reference Class 12
    3.2   Retrocausality 13
    Reference 13

4   **The Simulation Argument** 15
    4.1   Randomness and Reference Class 15
    4.2   Retrocausality 16
    4.3   Summary 17
    References 17

Part II   Dilemmas of Cooperation

5 Newcomb's Problem   21
    5.1   Preliminaries   21
    5.2   Three Problems, Four Theories   22
    5.3   Correlation   25
    5.4   Advisors   26
    5.5   Coherence   30
    5.6   The Perils of Incoherence   33
    References   34

6 The Open Box Game   35

7 The Hadron Collider Card Experiment   39
    7.1   Another Open Box Game   40
    References   44

Part III   Mystifying Envelopes

8 The Two-Envelopes Problem   47
    8.1   Opening the Envelope   48
    8.2   How to Play   55
    8.3   Summary   58
    References   58

9 Odds and Ends   59
    9.1   Doomsday   59
    9.2   The Betting Crowd   61
    9.3   Sims   61
    9.4   Newcomb's Problem   62
           9.4.1   Alternatives to Advisors   62
           9.4.2   The Consequence of Randomization   63
           9.4.3   Liebnitzian Lethargy   64
           9.4.4   Coherence Implies Stability   65
    9.5   Is the Card Game at all Feasible?   66
    9.6   Two Envelopes   67
           9.6.1   Additional Approaches?   67
           9.6.2   Causal Structure   70
           9.6.3   The Finite Two-Envelopes Game   71
           9.6.4   Ross's Theorem   72
    References   73

Contents

Epilogue: Anthropic Eden   75

Index   77

Introduction

The prevalence of paradoxes in probability theory can be judged against a subject famous for its paradoxes—mathematical logic, in which there are about half a dozen that are notable. (Russell's, Cantor's, Berry's, Burali-Forti's, Richard's, Grelling's and the Liar (Fraenkel et al. 1984) and (Heijenoort 1967)). Except for the Liar they all arose during the foundational period in the late nineteenth and early twentieth centuries. In probability theory (Szekely 1986) covers more than eighty paradoxes. (More advanced material of a paradoxical nature can be found in Romano and Siegel 1986). One can practically trace the whole history of probability theory in terms of paradoxes. And the probability paradoxes keep coming. Older paradoxes, settled and well understood, retain the capacity to instruct, astonish, and delight, but the paradoxes I have chosen remain controversial or are settled in a manner I believe to be incorrect.

We shall see time and again that the paradoxical nature of the materials drives analysts to resort to exotic, innovative, untested, fanciful, and otherwise dubious techniques. Proposed solutions are often as paradoxical as the original problems and only add to the mystification. The antidote is exclusive use of the near universally accepted elementary rules of probability theory. Five of these paradoxes have been widely discussed (Open Box and Card Experiment are new). Only two, the Betting Crowd and Two-Envelopes, are probability paradoxes in the narrowest sense. The others bear on questions of human nature, causality, human survival, cooperation, and controlling outcomes by actions. Each nevertheless hinges on the correct application of probability theory. The first three concern correct use of the concept of randomness, the second three the question of how probabilities should shape action. These problems are subject to definitive resolution in a manner that is routine in mathematics, but is in sharp contrast to the continual reassessment of old questions which is characteristic of philosophy. I claim that in every case correct application of probability concepts is the key to resolution of the paradox and that none of these applications are embroiled in foundational disputes. In other words, these are the kinds of probability problems whose solutions are independent of how one decides certain philosophical questions. The last claim rests uncomfortably with the fact that much of the dispution on these paradoxes is found in philosophical publications. Paradox itself invites philosophical treatment; after all, the greatest paradoxes from Zeno to Russell have been philosophical. My contention is that patient application of near universally accepted probability concepts is enough to resolve each paradox. Probability theory gives definite answers to well posed questions; in particular probability theory cannot give inconsistent answers to an unambiguous question. Ultimately there are no probability paradoxes, only probability *fallacies*. There are profound philosophical problems associated with probability concepts: determinism, the open future, subjectivity, induction—all of these and more have been brought into disputations over these paradoxes. They just happen to be irrelevant to this set of problems. In briefest terms there are three issues:

(1) The validity of a certain kind of Anthropic reasoning (Doomsday, Betting Crowd, Simulation)
(2) The role of dominance principles in decision making (Newcomb, Open Box, Card Experiment)
(3) The treatment of divergent expected value expressions (Two-Envelopes).

The required applications of probability theory are unambiguous for three reasons: (i) only *discrete* probability theory is needed. There is no call for the continua of real analysis. (ii) these problems are finitary in nature; the exception is the Two-Envelopes paradox but even here a finitary version can be developed (9.6.3). (iii) the solution to these paradoxes are independent of foundational questions such as Bayesianism versus frequentism. Since the subjective probabilities in these paradoxes can be cast in terms knowledge of frequencies, Bayesian and frequentist demonstrations run in parallel. In finitary probability problems dependant on observable frequencies, there is little space in which ambiguity and equivocation can hide.

The last four paradoxes concern the optimality of strategies. Treatment of such problems needs principles of decision making to link probabilities to evaluations and actions. Its desirable that such principles be intuitively obvious, e.g., the player should prefer the larger monetary payoff to the smaller one. I wish to promote three such principles that I take to be incontestable.

(1) **The Symmetrical Ignorance Principle**: symmetrical ignorance forestalls rational preference.
    Suppose everything a person knows about either of two options applies equally to both of them; then the person has no grounds on which to base a preference.
(2) **The Advisory Principle**: one should not recommend a strategy as optimal that fails to be optimal when the strategy in question is undertaken because of a recommendation.
    There may be initial doubt as to whether such strategies exist, but there is no denying that if there are strategies that do not work when recommended,

then someone who wants what's best for the player should not recommend such strategies.

(3) **The Coherence Principle**: problems in outcome alignment should be played in the same way.

Suppose two games are played in a series of rounds. On any given round if the games are played the same way, they always yield the same payoff. This may be a different payoff than last time this strategy was tried, but on any round, the payoffs are the same if the strategy used in both games is the same. The coherence principle states that if one game has a best strategy, that strategy is also best in the other game.

It is difficult to see any reason to deny validity to (1) or (2) as decision principles; moreover, there seems to be no motive for contesting them. Matters are different with (3). It is as obvious and compeling as the other two, but if (3) is granted (with support from (2)) a solution to Newcomb's problem ineluctably follows that most experts consider completely wrong. On the conceptually treacherous terrain of Newcomb's problem, one may be reluctant to accept even the most transparently obvious assertion, at least until one can examine its consequences. The reader who has reservations about the status of (3) can read Chap. 4 as placing upon proponents of any other solution the added burden of explaining why two problems that have exactly the same responses to player strategies should be treated differently.

The short and infrequent bursts of mathematics in this book, mostly in the chapter on the Two-Envelopes paradox, are of the sort that some find elementary and others impenetrable. I have tried to make it possible to bypass the mathematical parts with minimal loss, but they are needed for a full understanding of Two-Envelopes and Newcomb's problem. I wanted the argument to be compelling enough for dissenters to feel the need to locate a specific mistake. Both rigor and transparency can compel, but they are somewhat at odds with one another. I can only hope I've struck a good balance.

References

Fraenkel, Abraham A., Yehoshua Bar-Hillel, and Azriel Levy (1984) *Foundations of set theory*, North Holland, Amsterdam.

Heijenoort, Jean van (1967) *From Frege to Gödel*, Harvard University, Cambridge.

Romano, Joseph P., and Andrew F. Siegel (1986) *Counterexamples in probability and statistics*, Wadsworth & Brooks/Cole, Monterey.

Szekely, Gabor J. (1986) *Paradoxes in probability theory and mathematical statistics*, D. Reidel, Dordrecht.

Chapter 1
Seven Paradoxes

A Summary of the Paradoxes

1.1 The Doomsday Argument

How long will the human race survive? The Doomsday argument has been proposed not as paradox, but as a serious proposal about the prospects of human survival. It is paradoxical because elementary probability reasoning is used to go from a mundane fact to a truly extraordinary conclusion. As with any probability paradox it is a question of locating the fallacy.

Suppose a ticket is drawn from one of two lotteries, a small lottery with ten tickets, numbered 1 through 10, and a large lottery with a thousand tickets, numbered 1 through 1,000. A coin is tossed to determine from which lottery a random ticket is to be drawn. You are informed that the number drawn was 7, but not which lottery it came from. Although it could have been drawn from the large lottery, 7 is much more likely to have come from the small lottery. Bayes' Theorem can be used to show that the probability that the drawing was from the smaller lottery has shifted from 0.5 to nearly 0.99 (see Sect. 9.1). In the Doomsday argument popularized by John Leslie (Leslie 1990, 1992) an analogue of this reasoning is employed to make discoveries about the fate of the human race. This has stirred up considerable controversy. Leslie's argument, which he attributes to Brandon Carter, suggests the end of the human race may be much closer than we generally suppose, even in our current mode of ecological pessimism. The argument can be summarized as follows: among all people who have ever lived, our current birth-order rank is something like 60 billion. If humanity is to continue as long as we usually suppose, then we shall have a quite low rank among all who ever live. According to Leslie, this should be considered unlikely; it is more probable that we have middling rank among all who ever live, in which case doomsday will be much sooner than we generally suppose. In effect, one's birth is treated as though it were a random drawing from a lottery of unknown size, consisting of all humans who ever live. One's own rank in the drawing is used to estimate the size of the entire lottery pool. This suggests that the entire human lottery is not large relative to our rank, i.e., doomsday is likely to be relatively soon.

W. Eckhardt, *Paradoxes in Probability Theory*, SpringerBriefs in Philosophy, DOI: 10.1007/978-94-007-5140-8_1, © The Author(s) 2013

1.2 The Betting Crowd

The Shooting Room Paradox was devised by John Leslie to justify his Doomsday argument. In fact it does the opposite, serving to reveal the fallacious nature of Doomsday reasoning. In the original formulation, losing players are shot, but this gruesomeness distracts from the true paradox. I have replaced the jarring and irrelevant human carnage with betting. I call this the **Betting Crowd**: successive groups of individuals are brought together and are required to make the same wager; betting $100 that the "House," with fair dice, rolls anything but double sixes. Whenever the crowd wins its bets, ten times as many people as have played so far are recruited for the next round and a new roll of the dice. Once the House wins, the game series is over. So the House can truthfully announce before any games are played at least 90 % of all players will lose. The puzzle is that these bets appear to be both favorable and unfavorable, favorable because double sixes are rare, unfavorable because the great majority of players lose.

1.3 The Simulation Argument

Assuming technology advances, it should be possible to create sims (simulations that don't know they are simulations). Given enough time, sims should vastly outnumber real people. Since we cannot directly ascertain which we are, we're much more likely to be sims.

1.4 Newcomb's Problem

In **Newcomb's problem** the player is shown two boxes and given the choice of taking either the 1st box or both boxes. The first box is either empty or contains $1 million; the second box invariably contains $1000 and can be transparent. These boxes have been prepared by a resourceful entity, the predictor, a shadowy figure with an uncanny ability to judge which choice contestants will make. The predictor prepares the boxes according to the following rule: If the player is going to choose both boxes, leave the first box empty; if, however, the player is going to choose the 1st box only, place $1 million in that box. The predictor's abilities are such that it scores a consistent 90 percent in following the rule. The reason for either the predictor's successes or its failures is generally not given. What is best, to take one box or two? Either choice can be supported by a seemingly airtight argument:

(1) The one-boxer argument. The predictor may be mysterious, but the resultant monetary payoffs and their probabilities are precisely defined. The best course is to maximize expected value. Taking one-box leads to a payoff of $1 million

90 % of the time and zero dollars 10 % of the time, for an expected value of $900,000. Taking two-boxes leads to a payoff of one thousand dollars 90 % of the time and one of $1,001,000, 10 % of the time, for an expected value of $101,000. By this criterion the one-box strategy is far superior.

(2) The two-boxer argument. Let X represent the unknown contents of the 1st box. The player can choose one box and receive X, or choose two boxes and receive X + $1,000. Whatever the value of X, X + $1000 is always to be preferred, so the two boxers' strategy is consistently superior.

1.5 The Open Box Game

This is Newcomb's problem with both boxes open. In the first four paradoxes the conflict is evident. In the open box game it is difficult to find a paradox at all. The obvious answer seems to be the right answer; there seems to be no other side. One who has uncovered the paradox is well on the way to solving it. This matter is better left until Newcomb's Problem has been solved (Chap. 5).

1.6 The Hadron Collider Card Experiment

This paradox springs from a radical theory in physics that gives an astonishing explanation for various difficulties suffered by CERN' Large Hadron Collider, namely "something" in the future is trying by any means available to prevent the production of certain elementary particles called Higgs bosons. Operating from the future, unprecedented means are available.

1.7 The Two-Envelopes Paradox

There are two paradoxes called **Two-Envelopes**. Although related, they require different treatments.

The blind game: You're presented two envelopes, identical in appearance, and are informed that one envelope contains twice as much money as the other. (To conceal quantities, these can be checks.) You are randomly allotted one (we call this the 1st envelope) and then are offered the opportunity to switch envelopes. Since you have exactly the same knowledge concerning each envelope, the only reasonable conclusion is that you should value the envelopes equally, and hence be indifferent between switching or not switching. Alongside this is placed a paradoxical argument: suppose the 1st envelope contains F, then the other contains 2F half the time and F/2 half the time. So the other envelope is worth

$$(1/2)2F + (1/2)F/2 = 5F/4$$

an amount greater than F. This remarkable equation promises a 25 % average profit just from switching envelopes. However, this line of reasoning collides with itself: we can as easily denote the amount in the 2nd envelope by F, then symmetrical reasoning yields $(1/2)2F + (1/2)F/2 = 5F/4$ for the value of the 1st envelope, so by this account each envelope is worth 25 % more than the other.

The informed game: suppose the player is permitted to look into the 1st envelope before making the decision whether to switch. The amount in the 2nd envelope remains concealed until after the player's decision. Observing the amount in the 1st envelope breaks the symmetry of the game. Once the player learns F—the amount in the 1st envelope—she knows the 2nd envelope contains either F/2 or 2F. If, furthermore, she knows the probability distribution S from which the amounts were selected, she can determine the probabilities p and $1 - p$ of these alternatives. The expected value of the other envelope is

$$p(F/2) + (1 - p)(2F)$$

This is greater than F if an only if $(1 - p) > p/2$; that is, *it is favorable to switch when the probability of doubling is more than* 1/2 *the probability of halving*. One can find a distribution for which this relation holds for each value of F. For such distributions it would seem a winning policy to switch *irrespective of the value of F*. This policy can be fully implemented without opening the 1st envelope which brings us back to the first Two-Envelopes paradox and its insane advocacy for switching under symmetrical conditions.

References

Leslie, J. (1990). Is the end of the world nigh? *The Philosophical Quarterly, 40*(158), 65–72.
Leslie, J. (1992). Time and the anthropic principle. *Mind, 101*(403), 521–540.

Part I
Anthropic Fallacies

Chapter 2
DOOMSDAY!

The Doomsday argument concerns a question sure to arouse interest—the survival of the human race. It is rivaled only by the Simulation argument in attempting to address profound questions of human nature and destiny through elementary probability calculations. Proponents of the argument maintain that we are randomly selected from among all who ever live. This randomness makes it unlikely that we are among the earliest humans. A calculation demonstrates that if we are not among the earliest humans, then the human race has less time left than is usually allotted to it. Among all people who have ever lived, our current birth-order rank is something like 60 billion. If humanity is to continue as long as we usually suppose with the population sizes of the modern world,[1] then we shall have a quite low rank among all who ever live. This should be considered unlikely; it is more probable that we have an average rank among all who ever live, in which case doomsday will be much sooner than we generally suppose. Your birth rank (one plus the number of humans born before you) which can be estimated, corresponds to the number on the lottery ticket drawn (see 1.1). This rank is used to infer whether you are part of a big lottery (prolonged existence of the human race) or a small lottery (early doom). This suggests that the entire human lottery is not large relative to our rank, i.e., doomsday is likely to be relatively soon.

The ease with which the Doomsday argument extracts from a quite modest investment of current fact, substantial information about the remote future has provoked suspicion in most quarters; however, critiques that have not been snidely dismissive have tended to be as mystifying as the Doomsday argument itself. The Doomsday argument is a worthy paradox in that it is much easier to see the conclusion is unwarranted than to locate the fallacy. Bostrom (2002, p. 109) reports having heard more than one hundred attempted refutations of the Doomsday argument. The Babel of conflicting objections indirectly lends credence to the argument—more than a hundred attacks and still standing.

---

[1] It is the number of future individuals, not the amount of time to Doomsday, that allegedly regulates the sampling probabilities. Although we've had a fairly long history with low populations, the Doomsday argument indicates that assuming large populations, we have a short future.

W. Eckhardt, *Paradoxes in Probability Theory*, SpringerBriefs in Philosophy,
DOI: 10.1007/978-94-007-5140-8_2, © The Author(s) 2013

Proponents of the Doomsday argument base their reasoning on Bayes' Theorem, an uncontroversial proposition of probability theory (see 9.1). Use of Bayes' theorem should not be confused with Bayesianism, a controversial viewpoint on the nature of probability associated with subjectivism and the claim that correct statistical inference depends crucially on assessment of prior distributions. The most avid frequentist has no quarrel with Bayes' Theorem proper. Leslie blurs this distinction in a manner that allows him to portray the Doomsday argument as an application of a controversial doctrine (Bayesianism) to a straightforward fact of our existence (birth rank) with the suggestion that the argument, although a little dubious, mày, along with Bayesianism, turn out to be fundamentally sound. However, Bayes' Theorem is rigorously demonstrable (Feller 1968, p. 124); if the premises of the theorem are fulfilled, the conclusion follows with the force of logic. As with other rigorous results, the sticking point in applications is not whether the theorem is correct, but the extent to which the premises are fulfilled. The Doomsday argument should be ·seen as a straightforward application of an uncontroversial theorem (Bayes') to data produced from a highly questionable assumption (HR examined below). For this paradox and the next two, the fallacy is most easily traced in two stages: *randomness in reference class* and *retrocausality*.

2.1 Randomness in Reference Class

Bayes' Theorem permits the derivation of a conditional dependence from the converse dependence, in the presence of the right kind of background information. If our birth rank can tell us via Bayes' Theorem something about the likelihood of Doomsday, then it has to be because Doomsday can tell us something about our birth rank; there is no way around this. Doomsdayers make this connection by means of an assumption: the **Human Randomness Assumption** (HR): *We can validly consider our birth rank as generated by random or equiprobable sampling from the collection of all persons who ever live.* (I may have been the first to articulate this assumption (Eckhardt 1997, p. 248). Under HR the probability one has a given birth rank is inversely proportional to how many ranks exist. Chancing to have a birth rank as low as we do then makes it likely there are relatively few ranks available overall. Without HR the proposed application of Bayes' Theorem is trivial and fruitless, with HR the reasoning runs smoothly to its alarming conclusion. For a true lottery with an unknown quantity of tickets consecutively numbered, starting with 1, the random drawing of a ticket does indeed give us information about the probable size of the ticket pool. The issue in the Doomsday argument is whether such random lotteries constitute appropriate models, that is, whether HR is true.[2]

---

[2] In light of this Leslie's copious urn and lottery examples can all be seen to be question begging; each one assumes equiprobability of sampling, precisely what must be established to validate

If we are random[3] humans, are we also random primates? random vertebrates? random readers of English? Leslie proposes that a single human be taken as a random sample from the class of all humans *and* from the class of all mammals (Leslie 1993, p. 491). One cannot be random in more than one among such diverse classes. Suppose x were a random human *and* a random mammal. x is human, and hence a most unusual mammal; conversely a random mammal should not be a human except by the most freakish chance. How would you go about selecting x? Choose from among humans and it is not a random mammal, choose a random mammal and it is almost surely not a human; keep choosing random mammals until you get a human and you have spoiled the randomness of your mammal selection. The instruction to select an x that is random in both these classes is incoherent. Since different randomness reference classes yield different statistics and different conclusions for Doomsday reasoning, on what grounds choose humans as the unique class for the argument? Doomsdayers appear at times to take this choice for granted, at other times to suggest alternatives that contradict it, such as that the reference class should be all intelligent individuals including extraterrestrials or simulated people.

Examine how the allegedly random user of the Doomsday argument is selected. Self-selection is undertaken only after the Doomsday argument, or its central concept that we are randomly selected humans, is invented. HR amounts to the claim that we can select ourselves as random, on the basis of the invention of the Doomsday argument. Inventions are chancy but not purely random, e.g., they are closely related to the technological and cultural conditions of the time and are not equally likely to appear at one time as another. Since there is no reason to believe this invention occurs randomly, there is no reason to believe that self-selection dependent on the invention yields a random human.

2.2 Retrocausality

Even if the considerable problems of reference class were resolved, there looms a more serious obstacle. The HR assumption stipulates a quantitative relationship between the probability of having your birth rank and the number of people who

---

(Footnote 2 continued)
Doomsday reasoning. (In response to my objection to the assumption of random human sampling in the Doomsday argument, Leslie produced additional urn and lottery analogies presupposing random sampling (Leslie 1993)). Leslie seems to believe these matters can be settled through sheer weight of accumulated analogies. Although of possible pedagogic or heuristic value, such analogical reasoning can at best support only the preliminary stages of investigation, whereafter it becomes incumbent to find nonanalogical evidence or otherwise to investigate the validity of the analogies.

[3] I use **random** only to refer to an equiprobable sampling on a finite set, i.e., probability $1/N$ is assigned to each of N objects.

come *after* you. For the argument to be valid, the crucial sampling probability has to be based not only on how many were born before you but also how many are to be born after you. How it is possible in the selection of a random rank to give the appropriate weight to unborn members of the population? In presuming that unborn populations have somehow been factored into the current selection procedure the Doomsday reasoners tacitly presuppose retro-causal effects that is, the effects comes before the cause. Suppose some crucial event prevents a catastrophe in the year A.D. 2050 and doom is thereby delayed a thousand years. By Doomsday reasoning, the probability of having your birth rank is therefore lower than it would have been were humans to become instinct in 2050. The Doomsday argument tacitly requires that future events influence current ones. You have a *random* birth rank whose expected value consequently depends on the total number who ever live which itself depends on the number who come after you. Consider a standard lottery with numbered tickets to which higher numbers are subsequently added; this addition raises the average value of a number drawn. Similarly, for HR to hold, all humans, even unborn ones, must be accorded their appropriate weight in the selection of a random birth rank. The Doomsday argument hinges on correlation[4] between your allegedly random birth rank and the size of future populations because a protracted future tends to boost random birth ranks, and a short future tends to depress them. Only if future populations exert this kind of influence on current births can knowledge of the relative lowness of your birth rank be so informative about the distant future. Such retrocausal influence is not inconceivable; however, the need for it further compromises the already tattered plausibility of the Doomsday argument.

References

Bostrom, Nick. (2002). *Anthropic bias; observation selection effects in science and philosophy*. New York: Routledge.

Eckhardt, William. (1997). A shooting-room view of doomsday. *The Journal of Philosophy, 97*, 244–259.

Feller, William. (1968). *An introduction to probability theory and its applications* (Vol. 1). New York: Wiley.

Leslie, John (1993) Doom and probabilities *Mind 102*(407), 489–491.

---

[4] By **correlation** we mean simply "lack of independence". The term "dependence" may seem more apt, but dependence is often asymmetric whereas independence and correlation are symmetric. In this framework **positive correlation** between events X and Y means that $P(XY) > P(X)P(Y)$ and negative correlation that $P(XY) < P(X)P(Y)$. (This kind of correlation is sometimes called stochastic or probabilistic correlation.).

Chapter 3
The Betting Crowd

Of the seven paradoxes the Betting Crowd is the easiest to penetrate. Therein lies
its value; analysis of the fallacious aspects of the Betting Crowd can greatly help in
negotiating the more treacherous terrain of the Doomsday and Simulation arguments. The non-murderous version we examine concerns bets on dice, making
fallacies easier to spot. The Betting Crowd game consists of one or more rounds.
For each round a certain number of players enter a region and they each bet even
money *against* double sixes on a single roll of the dice (so they all win or lose
together). If the players win, they leave the region and a number of new players are
brought in that equals ten times the number of players that have won so far. The
dice are rolled again. The rounds continue until the house wins on double sixes at
which point the game is over. This guarantees that 90 % of all those who play lose.
Two trains of thought collide: (↑) since double sixes occur less than 3 % of the
time and a player stands to win about 97 % of the time, the bet is highly favorable;
(↓) since 90 % of all players are destined to lose, the bet is highly unfavorable.

Before the dice are rolled, how should a player in the crowd regard her prospects? The house has a scheme that assures it a final profit because eventually
double sixes turn up. For (↓) to be true the future profitability of this scheme would
need to seep back into the current roll of the dice—a retrocausal influence. Neither
the house's strategy, nor the fortunes of other players, nor the play in other rounds,
can alter the favorability of betting against double sixes in a roll of fair dice. The
House triumphs not by increasing the likelihood of double sixes but by the sheer
number of players that lose when double sixes occur. It is the likelihood of double
sixes not the contrived arithmetic of the room populations that matters to the
individual player placing a bet. We conclude (↑) is undoubtedly correct, and (↓) is
wholly fallacious.

There is solid dependence among outcomes for the players in a single round—
they all win or they all lose together. If five players lose independently at a given
kind of wager, that may constitute, in a loose sense, five reasons not to play, but if

W. Eckhardt, *Paradoxes in Probability Theory*, SpringerBriefs in Philosophy,
DOI: 10.1007/978-94-007-5140-8_3, © The Author(s) 2013

a million players all lose the same wager at once, that constitutes, in the above sense, *one* reason not to play.[1] It has long been known that by successively increasing bet size in a sequence of unfavorable bets one can theoretically obtain winning results (e.g., (Epstein 1977, pp. 52–56) this is the basis of various infamous doubling systems in Roulette and other games. In the Betting Crowd game it is the House that carries out the "multiplicative" betting scheme. A player ought to reason thus: 90 % of all players will lose, but I have less than a 3 % chance of belonging to that losing majority. This is no paradox; each player is prospectively likely to be in the minority, since he or she is prospectively likely to win and winning itself causes there to be enough subsequent players to guarantee the winner is in the minority.

Suppose one player has the opportunity to play in hundreds of different Betting Crowd series. Since rolls of the dice are independent of one another, she would lose about one game in 36 and win in the other 35. There would be statistical variation in these results, but how could the teeming multitudes of losing players make her more likely to lose? Such freakish behavior could only mean the dice were blatantly rigged. With the fair dice specified in the problem she would with high probability *win* close to 97 % of her games, thereby netting many thousands of dollars playing games which are unfavorable to the player according to those who succumb to the paradox.

3.1 Randomness and Reference Class

Before the dice are rolled there is no appropriate reference class for the player. Her prospects are governed by the dice probabilities which are not reflected in any group in the story. (One might artificially pick out a group with the required proportions, but this group would play no role in the game.) Randomness in a reference class is an unwieldy concept in open ended processes such as human survival, Betting Crowd series, or the sim production of the next paradox, at least until the process is completed.

In the Betting Crowd story the final frequencies are known beforehand with unusual precision. This can make it appear that the determinate statistics of the final pool are already operative when the dice are rolled. Before the role of the dice the player has only about a 3 % chance of belonging to the 90 % majority. What lends an air of paradox is that an *unlikely* event makes a group of players *typical*. The player should not consider herself a typical member of the population unless

---

[1] It has been remarked that an insurance company would go broke insuring all the players in the [Shooting Room] as though they had a 97 % chance of winning. And so might any insurance company that treats highly dependent events as though independent—a diversified company that finds it rarely receives simultaneous flood claims and decides to insure everyone living on the banks of the Mississippi.

she receives double sixes. This means the player should not consider herself random until the game series is over.

3.2 Retrocausality

The Betting Crowd paradox rests on an unconvincing appeal to a retrocausal influence: the sheer quantity of the eventual losers is said to dampen the player's prospects in the present. Of the seven only the Betting Crowd paradox harbors so blatant a fallacy; its importance resides in its mimicry of the more confounding fallacies of the Doomsday and Simulation arguments.

Summary: There are parallels between Doomsday reasoning and fallacious reasoning in the Betting Crowd. In both the assumption of the user's randomness is needed for the argument to proceed. In both this randomness implies an influence from the future. In the Betting Crowd this kind of thinking can be made to yield absurd and contradictory results; in the case of the Doomsday argument, the elusiveness and ambiguity of human destiny help to conceal the argument's invalidity.

Reference

Epstein, R. A. (1977). *The theory of gambling and statistical logic*. New York: Academic.

Chapter 4
The Simulation Argument

A **sim** is a conscious simulation that does not know it is a simulation. Current
enthusiasm notwithstanding, it is a distinct possibility that creating this good a sim
is either unfeasible or impossible in which case the Simulation argument does not
get started. Bostrom (2003) discusses possible developmental paths a technological civilization might take that would forestall the creation of superabundant sims.
Since I'm interested in the validity of the underlying argument, not the question of
what technologies might arise, I make technological and sociological assumptions
conducive to the argument, namely, that it is ultimately easy to proliferate sims,
and that eventually sims greatly outnumber the unsimulated.

The Simulation argument has a disarming simplicity. If overall there are many
more sims than real people, and you cannot tell which you are, then you are most
likely a sim. There is no tricky reasoning about birth rank as in the Doomsday
argument, yet both arguments conclude it is later than we think, the first by
shortening humanity's future, the second by lengthening humanity's past.

4.1 Randomness and Reference Class

The quandaries of reference class in the Doomsday argument transfer to this case: if
simulated, are you random among human sims? hominid sims? conscious sims? The
Simulation argument requires an *extended randomness* (ER) assumption for which
the reference class commingles both real people and sims. This heterogeneity makes
it harder to justify a seamless equiprobability spread evenly throughout the class as
randomness demands. Moreover, HR and ER are incompatible. If you are a sim,
then your birth rank is much higher than 60 billion. Writing in 1993, in advance of
the Simulation argument, I employed the now quaint idea of human brains inside of
robots to suggest that Sim-like considerations undercut the Doomsday argument:
"Suppose in one hundred years people stop reproducing in the way that is currently
customary and for the next million years the human race consists of human brains

W. Eckhardt, *Paradoxes in Probability Theory*, SpringerBriefs in Philosophy,
DOI: 10.1007/978-94-007-5140-8_4, © The Author(s) 2013

inside of robots. Is this a confirmation of the doomsday reasoning because among all flesh and blood humans we then have average rank? Or is it a disconfirmation demonstrating the need for some reason as to why we chanced to be born so early in the history of the race that we are not brains inside of robots?" (Eckhardt 1993, p. 484).

4.2 Retrocausality

Suppose enough time has passed since the advent of sim technology that there are *now* vastly more sims than real people. It would then be reasonable to conclude that a random birth is almost surely a sim, just as it is reasonable to conclude that a random birth is more likely to be in Asia than Iceland. How do we get from the concession that at some point in history sims definitively come to dominate the population to the conclusion that we now live at such a time and are sims? Only by means of ER: if you are random among all sims and people and if nearly all of these are sims, then you are almost surely a sim. According to this reasoning, every sim, even an as yet uncreated one, increases the probability you are a sim.

The Doomsday argument mimics predictive inference since *current* knowledge of birth rank is used to make inferences about *future* doom, but beneath the surface it runs in the opposite direction: later doom induces higher average birth ranks. With the Simulation argument the temporal inversion is more explicit: the case for future sim creation leads to inferences of past and present sim creation. Proponents of the Doomsday or Simulation argument must not only overcome the considerable case against the existence of any kind of (observable) time-reversed causation (see Eckhardt 2006), but must also find specific justification for the peculiar effect unborn populations are claimed to have on one's birth rank.

In the Betting Crowd once it is realized the bets should be judged by same rules as any bets on dice, it becomes an excellent proving ground for the other two paradoxes. Consider the results of applying the "Anthropic" reasoning of the Doomsday and Simulation arguments to the Betting Crowd.

	Doomsday	Simulation
	I am random among humans, therefore my birth rank is average among humans.	There are sims and real people, but I do not know which I am. Ultimately there are vastly more sims than real people, so I'm more likely to be a sim.
Betting Crowd	I am random among players and nine out of ten players lose, so I'll probably lose.	There will be winners and losers, but I do not know to which I belong. Ultimately there are many more losers than winners, so I'm more likely to lose

By this reasoning greater future populations increase the probability you have a higher birth rank; greater populations of future sims increase the probability you are a sim, and greater populations of eventual losers in the game increase the probability you will lose. It does not bode well for its application to nebulous questions that this kind of reasoning fails outright when applied to a clear case.

4.3 Summary

Two arguments, one that the end of humanity is closer than we might otherwise expect and another that our world is most likely a simulation, are quite similar despite starkly different conclusions. Each depends crucially on a problematic kind of human "randomness", each suffers from gross indeterminacy as to the correct group displaying this randomness, and each relies on an implied influence of the future upon the present.

References

Bostrom, Nick. (2003). Are you living in a computer simulation? *Philosophical Quarterly*, *53*(211), 243–255.

Eckhardt, William (1993) Probability and the doomsday argument *Mind*, *102*(407), 483–488 (available at www.WilliamEckhardt.com).

Eckhardt, William (2006) Causal time asymmetry *Studies in history and philosophy of modern physics* 37, 439–466 (available at www.WilliamEckhardt.com)

Part II
Dilemmas of Cooperation

Chapter 5
Newcomb's Problem

5.1 Preliminaries

The first three paradoxes revolved around questions of what would be factually true, given certain assumptions. The rest concern questions of what someone called the **player** *should* do given certain situations,[1] thereby entering the realm of **decision theory**. To guard the claim that these paradoxes depend on probability concepts, the possible meanings of "should" must be severely curtailed. This can be accomplished by stipulating a few intuitively obvious rules such as that given a choice between two options, the player *should* take the one of greater value, and also by stipulating the values assigned to relevant elements in the problem.

In an oversimplification of the history of empirical science we might say that experiments inspire theories, and theories suggest experiments. In decision theory, which purports to be the science of decision making, experimentation has a different status.[2] Although there are significant disagreements among variant decision theories, the question of experimental investigation as a means of settling theoretical disputes does not arise. The specifications of a typical decision problem are such that it is easy to predict average results in a long series of trials. It would indeed be pointless to resort to experimental trials when the results of such trials are easily and reliably predicted. What is unexpected is that all current contenders for the "correct" decision theory at one point or another fly in the face of what such experimental trials would reveal. There exist a variety of arguments both for and against one-boxing, but in keeping with the design of this book, I search for an incontrovertible argument. (Of course it will be controverted.)

---

[1] The Betting Crowd is superficially of this form (note the players aren't given a choice) but the betting is as eliminable as the shooting. The core paradox rests on a probability question: on any round before the dice are rolled, does the player have a higher probability of winning or losing?
[2] We are speaking exclusively of *normative* decision theory which seeks the rational or optimal decision; *descriptive* decision theory, which seeks to understand human decision making behavior, is an experimental science in the standard sense.

W. Eckhardt, *Paradoxes in Probability Theory*, SpringerBriefs in Philosophy, DOI: 10.1007/978-94-007-5140-8_5, © The Author(s) 2013

We narrow attention to problems of a quite specific form: a player faces one of two possible situations and can make a binary choice. If these factors together determine the result, there are four possible outcomes. We use C and D as labels for the binary choice offered to the player. In an important class of games— cooperation games—C stands for "cooperation" and D for "defection" The common structure facilitates comparisons. We use "problem" and "game" interchangeably; we also make no distinction among "action", "choice", and "decision". "Correlation" and "independence" are probabilistic, unless they are specifically identified as causal.

5.2 Three Problems, Four Theories

The question we address is how to factor into decision making various kinds of correlations between actions and outcomes. We review three storylike problems that have proven crucial to this question.

I repeat the description of **Newcomb's problem**: the player is shown two boxes and given the choice of taking either the 1st box or both boxes. The first box is either empty or contains 1 million[3]; the second box invariably contains 1,000 and can be transparent. These boxes have been prepared by a resourceful entity, the predictor, a shadowy figure with an uncanny ability to judge which choice con-testants will make. The predictor prepares the boxes according to the following rule: If the player is going to choose both boxes, leave the first box empty; if, however, the player is going to choose the 1st box only, place 1 million in that box. The predictor's abilities are such that it scores a consistent 90 % in following the rule. The reason for either the predictor's successes or its failures is generally not given. What is best, to take one box or two? Either choice can be supported by a seemingly airtight argument:

(1) The one-boxer argument. The predictor may be mysterious, but the resultant monetary payoffs and their probabilities are precisely defined. The best course is to maximize expected value. Taking one-box leads to a payoff of 1 million 90 % of the time and zero dollars 10 % of the time, for an expected value of 900,000. Taking two-boxes leads to a payoff of one thousand 90 % of the time and one of 1,001,000, 10 % of the time, for an expected value of $101,000. By this criterion the one-box strategy is far superior.

(2) The two-boxer argument. Let X represent the unknown contents of the 1st box. The player can choose one box and receive X, or choose two boxes and receive X + 1,000. Whatever the value of X, X + 1,000 is always to be preferred, so the two boxers' strategy is consistently superior.

---

[3] I omit the dollar signs from monetary amounts and expected values; the numbers can then be interpreted as money or as units of utility. In the latter case "expected value" should be replaced with "expected utility".

Since the problem's publication, it has generated a sizeable literature. Abstruse elaborations and amendments to decision theory have been proposed in an effort to arrive at the 'correct' response to Newcomb's problem; free will, the open future and the nature of mind, have been brought into the fray. Most effort on this subject has been to bring decision theories to a two-box conclusion. This is the primary motivation for the sundry causal decision theories and of the various ramified or diachronic reformulations of evidential decision theory, proposed in the late 20th century.

The **Prisoner's Dilemma** has been the focus of vast amounts of discussion and analysis. It's realism contrasts favorably with the fantastical aspects of Newcomb's game and the Solomon story. In this much scrutinized predicament, two prisoners have to decide individually and independently—that is, without any communication between them—whether to confess or not. The consequences, known to both parties, are as follows: (1) if both confess, both serve 5 years. (2) if neither confess, both serve 1 month (on a technicality). (3) if one confesses and one doesn't, the first goes free and the second serves 10 years (the first has turned "states evidence" against the second). The only way a prisoner can effectively guard against disastrous consequences of being the only one not to confess is to confess himself. The situation in which both prisoners confess has a perverse kind of optimality in that either prisoner does much worse by unilaterally changing strategy, but each would do much better if both changed strategy; both prisoners are stuck in an "optimum" that falls far short of what they could obtain if there were some mechanism to assure cooperation. There has arisen a firm body of opinion that defection is the appropriate strategy in this game, even though mutual cooperation offers a much more favorable joint payoff. However, there have been sundry arguments advanced for cooperation in the Prisoner's Dilemma, most based on alleged correlations between the prisoners' behavior. To emphasize the role of player correlation we define the **tribal PD**, a prisoner's dilemma in which players are drawn from a population made up of two tribes that are distrustful of one another. It is found that prisoners drawn from the same tribe cooperate 90 % of the time, whereas prisoners from different tribes defect 90 % of the time. Prisoners from the same tribe correlate by cooperating more, and prisoners from different tribes correlate by defecting more.

Less attention has been paid to the **Solomon story**, a parody of Newcomb's problem that allegedly reveals the folly of one-boxing. Solomon is an ancient monarch vaguely reminiscent of the Israelite King. (Every part of this story is Biblically inaccurate.) He is pondering whether to summon Bathsheba, another man's wife. But Solomon is also fully informed as to the peculiar connection between his choice in this matter and the likelihood of his eventually suffering a successful revolt: 'Kings have two basic personality types, charismatic and uncharismatic. A king's degree of charisma depends on his genetic make-up and early childhood experiences, and cannot be changed in adulthood. Now charismatic kings tend to act justly and uncharismatic kings unjustly. Successful revolts against charismatic kings are rare, whereas successful revolts against uncharismatic kings are frequent. Unjust acts themselves, though, do not cause successful revolts... Solomon does not know whether or not he is charismatic; he does know that it is unjust to send for another man's wife.' (Gibbard and Harper 1978) It is intuitively evident that Solomon's restraining himself from summoning Bathsheba does not lessen the likelihood of revolt. Since the problem does not include the negative moral value of this action, it would seem the optimal course for Solomon is to summon.

Newcomb's problem has fractured decision theory into a host of warring parties and engendered formulations that are complicated, inelegant, and, I would venture, incorrect. These newer theories share a common trait: they are all more or less self-conscious attempts to secure a two-boxer resolution to Newcomb's problem. Most disagreement between experts on this subject concerns the correct way to reach this conclusion. This entire edifice, its concordances and its disputes, are vulnerable to the possibility that two-boxing is the wrong way to play.

Four kinds of decision theory are relevant to Newcomb's problem.

(1) **Evidential decision theory (E)**. (Jeffery 1965). In this theory, what matters is the degree to which one's action modifies the probabilities of outcomes. The recommendation is always to take the action with the greatest expected value.
(2) **Causal decision theories (C)**. This refers to a group of essentially equivalent formulations (Gibbard and Harper 1978; Lewis 1981; Skyrms 1982) all based on the principle that it is exclusively the causal consequences of one's acts that are relevant to decision. The only relevant probabilities are the probabilities that a given action would cause a given outcome. In Newcomb's problem, the causal decision theorist reasons as follows: the player's choice can have no possible causal influence on the box contents; if the 1st box contains X, both boxes together contain X + \$1,000, which is always to be preferred to X alone. This is the classic two-boxing rationale in Newcomb's game.
(3) **Reformed evidential decision theories**. This refers to a heterogeneous group of formulations, e.g., (Eells 1982, 1984; Jeffery 1983 and Kyburg 1980) that seek to obtain most of the recommendations of causal decision theory by evidential means.
(4) **Coherent decision theory (D)**, the formulation proposed in this article.

The case for rejecting the evidential theory is impressive. It would have you not confess in the Prisoner's Dilemma, evidently in a misguided attempt to influence the other prisoner, who will not even know of your decision, and would have Solomon not summon Bathsheba in what is evidently an even more misguided attempt to alter his own past. The Solomon story, a parody of Newcomb's problem, was explicitly designed to foil the evidential theory and seems to have succeeded in this task.

There have been a variety of arguments and counterarguments about the right way to play in Newcomb's problem, mostly a contest between an inchoate intuition that one-boxing provides a great opportunity and a closed dogma that one-boxing is for fools. Instead of adding to this din (I do that in Sect. 9.4) I outline a procedure for solving all decision problem of a certain kind including Newcomb's problem. The procedure is somewhat roundabout: virtual experiments are proposed in which a game under investigation is repeatedly played under controlled conditions. Certain flawed experimental designs must be avoided. The results of such experiments are calculated, and from this it is deduced how the game should be played. This procedure is also used to treat the next two paradoxes.

5.3 Correlation

Probabilistic **independence** of X and Y means $P(XY) = P(X)P(Y)$. Suppose neither $P(X)$ nor $P(Y)$ is zero. Then $P(Y|X) = P(XY)/P(Y) = P(X)$; similarly $P(Y|X) = P(Y)$. These express that Y does not change $P(X)$ and X does not change $P(Y)$. IF X and Y are not independent, then they **correlate**. Proponents of a probabilistic approach to causality speak of **screening off**. It turns out this is another name for **conditional independence**. Basically, Z screens X off from Y, if correlation between X and Y originates in Z or passes through Z.[4] There are essentially three sources of correlation:

(1) **Causal correlation**. This refers only to the correlation of a cause with its effect. A cause is screened off from its effect by any intermediate stage of the causal process. For example, in most cases, lighting the end of the fuse causes a firecracker to explode; the lighting correlates with the exploding. If the middle of the fuse is burning, the firecracker is equally likely to explode whether the end of the fuse was lighted or not (e.g., the middle may be burning because the end had been lighted or because the middle had been lighted). The burning of the middle of the fuse screens off the explosion from the lighting of the end.

(2) **Common-cause correlation** refers to the correlation between two effects of one cause. This is usually called **causally spurious correlation** but this correlation is causal in origin and not at all spurious. The common cause of effects screens off the causally spurious correlation between the effects. A famous but somewhat outdated example is that lung cancer correlates with yellow stained fingers. (Cigarette smoking used to leave yellow stains on certain fingers.) A person with yellow stained fingers is more likely to have lung cancer than a person with unstained fingers. This correlation owes to a common cause of both—cigarette smoking. Conditional on cigarette smoking, these no longer correlate, i.e., conditional on smoking, the probability of lung cancer is the same with or without yellow fingers; conditional on not smoking, the probability of lung cancer is the same with or without yellow fingers. This is no

---

[4] The usual definition of screening off is $P(X|YZ) = P(X|Z)$. (Eells 1991 p. 223) This can be rewritten $\frac{P(XYZ)}{P(YZ)} = \frac{P(XZ)}{P(Z)}$. Multiply both sides by $P(YZ)/P(Z)$. Then $\frac{P(XYZ)}{P(Z)} = \frac{P(XZ)}{P(Z)}\frac{P(YZ)}{P(Z)}$ which is $P(XY|Z) = P(X|Z)P(Y|Z)$ the definition of independence conditional on Z.

paradox since independence is conditional. Are stork populations independent of car sales? Yes, *conditional* on being on Earth. Drop this tacit condition and you find both storks and cars on Earth but neither elsewhere in the Universe. The correlation is spectacular! Arguably all events correlate because of common cause correlation from the Big Bang; then, in our world at least, independence is always conditional.

Causal and common cause correlations can comingle; in Newcomb's problem the correlation between boxing choice and receiving the thousand is causal that between boxing choice and receiving the million is common cause.

(3) **Chance**. Randomness can take on the appearance of correlation, e.g., coincidence. This is a challenge for the statistician, but in typical decision problems relevant probabilities (and at times irrelevant ones!) are precisely specified so correlations can be calculated; one need not be concerned about how to treat experimental flukes.

The key problems all turn on the status of common cause correlations. In Newcomb's problem it is presumably prior traits of the player that cause both the player's decision and the predictor's forecast. In the Prisoner's Dilemma, any correlation between the prisoners' decisions can only arise from factors such as culture or human nature that influence both of them prior to their seclusion. In the Solomon story charisma is the common cause both of summoning and of revolt.

A decision theory T can be characterized in terms of its *relevance set* $R(T)$ which describes the kinds of act-outcome correlations the theory accepts as relevant to decision making. Familiar examples are $R(C)$ the causal correlations, and $R(E)$, the evidential correlations. A decision theory T is **acceptable** if $R(C) \subseteq R(T) \subseteq R(E)$. No one would endorse a decision theory that ignored causal consequences of the outcome of the recommended course of action, so causal act-outcome correlations need to be in $R(T)$. At the other extreme if an event is uncorrelated with a choice, i.e., performing the chosen action neither raises nor lowers the probability of the event, then the event does not count as an outcome of the action, and there is no reason to consider the event in assessing the value of the action. Therefore nothing outside of $R(E)$ belongs in a relevance set. If $R(C) \not\subseteq R(T)$, T is negligent; if $R(T) \not\subseteq R(E)$, T is superstitious and irrational. From this point on "decision theory" means acceptable decision theory.

5.4 Advisors

A decision theory's recommendations should accord with well designed experiments for testing such recommendations. It's necessary to specify what constitutes a well designed experiment in this context and to determine what can be inferred from such experiments. In this section virtual experiments are defined for any problem of the designated form; in the next it's shown how the optimal choice in a problem can be determined from the experimental results. The experiments are organized as a sequence of trials. Each entry in a **player sequence** consists of a player, a choice, and an outcome. For each trial a player, who is randomly drawn from a large population, makes a choice in the problem and receives an outcome. It greatly facilitates analysis of a player sequence for the trials to be independent and identically distributed; to this end we make the following stipulations: the setup, including the predictor, does not improve or degrade as matters proceed; players play only once; they are sampled with replacement; if a player is resampled, he is associated with the same choice and outcome as before.

Advisory Principle: a decision theory should not recommend actions as optimal that fail to be optimal when the actions in question are performed because of recommendations.

Violation of this principle opens the door to bad recommendations. If there are indeed strategies that lose their "optimality" when recommended, it's transparently obvious that such strategies ought not to be recommended. I take it to be a fundamental principle of decision theory. For a recommendation to be judged optimal, the alternative strategies have to be assessed in terms of their values as recommendations. A well designed experiment then needs an apparatus for making recommendations to players. I follow the natural procedure of supposing the recommendations come from persons acting as advisors (alternatives are reviewed in Sect. 9.4). Each entry in an **advisor sequence** consists of an advisor, a player, a choice and an outcome. Each player is randomly paired with an advisor, drawn from a large populations of advisors. The pairings are known to all parties. We attribute the following characteristics to advisors:

1. advisors participate only once and are sampled with replacement; an advisor who happens to be resampled may be randomly associated with a new player, but the advisor brings with her the same recommendation and outcome as before.
2. an advisor has to recommend what she truly believes to be best for the player (no undecided advisors).
3. it has to be the recommendation of an acceptable decision theory.
4. a player has to follow this advice; the player has no influence at all on the advisor.

((1) assures that trials originating from advisor sequences are independent and identically distributed. (2) and (3) assure that advisors play a role analogous to that of decision theory whose purpose is to determine what is best for the player or decision maker without mendacity or malevolence. As for (4), to rule otherwise would be to unleash unpredictable advisor-player negotiations that would have no bearing on the original problems.)

Advisors model the role of a decision theory; they recommend what they truly take to be in the player's best interests, and the advised player follows a recommendation rather than his own inclinations. Advisors make recommendations like a decision theory and make decisions like a player. To prevent violation of the advocacy principle, well designed experiments can employ advisor sequences.

Since advisors are randomly paired with players, advisors induce random variation in the players' choices. The resulting the act-outcome correlations measure the extent to which outcomes can be *manipulated* through arbitrary alterations of the choices made.

A **causal problem** is one in which the choice-outcome correlations owe exclusively to the causal influence of choice upon outcome[5]; in this case a causal chain extends from advisor to player to choice to outcome, and the choice screens off the advisor or player from the outcome. A **deterministic problem** is one in which choice-outcome correlations are perfect. In this case the outcome of a given choice is always the same. A problem that is not deterministic is **probabilistic**. Regarding "causal" and "deterministic" neither implies the other: CA(0.9, 0.9) is causal and probabilistic, NP(1, 1) is deterministic and non-causal (symbols defined below).

An originally deterministic problem can be rendered probabilistic by means of an **external probability source**. This is a Bernoulli process $B(p)$ that is consulted on each trial. A fraction p of the time $B(p)$ registers a success, and the deterministic problem proceeds as usual; a fraction $(1 - p)$ of the time $B(p)$ registers a failure, and the results of the deterministic process are switched, e.g., an empty 1st box is filled or a full 1st box is emptied. A deterministic problem can as well be externally sourced by two Bernoulli processes, $B(p_1)$ for C-players (cooperators) and $B(p_2)$ for D-players (defectors).

Let $NP(q_1, q_2)$ be a Newcomb's game in which the predictor's success rate is q_1 for one-boxers and q_2 for two-boxers. Let $CA(p_1, p_2)$ be a purely causal problem which uses $B(p_1)$ and $B(p_2)$ as external probability sources in the following way: the player chooses one box or two; a clerk records this decision and places one million in the 1st box if the player chose one box, and leaves the box empty if the player chose two boxes. The only snag is that the clerk first consults one of two Bernoulli processes. $B(p_1)$ for one-boxers, $B(p_2)$ for two-boxers. If the Bernoulli process in question registers a success, matters proceed as above. If the Bernoulli process registers a failure, the allocation is switched: empty box for one-boxers, the million in the box for two-boxers. For high values of p_1, the player's one-boxing *causes* the box to have the million even though this does not work a small percentage of the time. (The marksman *causes* the bullet to hit the target; this does not mean she never misses.) We use NP and CA for NP(0.9, 0.9) and CA(0.9, 0.9) respectively.

Two problems are **outcome aligned** or **in outcome alignment** if on any trial in which advisors direct the same choice in each problem, the players receive identical outcomes. Two problems are **outcome alignable** if they can be brought into outcome alignment. Problems that vary independently of one another cannot

---

[5] Jeffery (1983, p. 20) speaks of "the central heartland of decision theory, where probabilities conditionally on acts are expectations of their influence on relevant conditions". This is the territory of causal problems where all decision theories work together in harmony.

be in outcome alignment. There are three ways in which outcome alignment is possible.

1. The problems are deterministic. For instance NP(1, 1) and CA(1, 1) are in outcome alignment even though one is non-causal and the other causal. NP(0, 0) and CA(0, 0) are also in outcome alignment.
2. The problems begin as deterministic but share an external probability source. For instance NP(1, 1) and CA(1, 1) can share a B(0.9) probability source so that both problems have their outcomes switched on the same trials. The resulting problems which resemble NP and CA are outcome aligned.
3. One problem is probabilistic and the other begins as deterministic. The probabilistic problem acts as the external probability source of the deterministic problem, bringing them into outcome alignment. For instance NP and CA can be brought into outcome alignment if NP is played first, and on those trials in which the predictor errs the results are switched in CA.

How should outcome aligned problems be played separately, i.e., with separate players, each having no monetary interest in the other game? (The correlation between the games complicates the question of how the games should be played together, i.e., with the same player in both games.) For (1) it is clear that neither problem has any effect on the other. Outcome alignment results from the internal operation of each problem. The fact of outcome alignment does not alter the way in which either problem should be played whether separately or together. In (2) once $B(p_1)$ or $B(p_2)$ is consulted, each problem proceeds along separate channels. Choices or outcomes in one problem have no effect on outcomes in the other. Outcome alignment has no effect on how the problems should be played whether together or separately. In (3) the probabilistic problem exerts an influence on the deterministic problem; however, this does not affect how the games should be played separately. The source arrangement has no effect on how the probabilistic problem should be played since it is completed before the other problem begins. As for the other problem it does not matter to optimal play whether the Bernoulli processes that source it arises from spinners, random number tables, or a Newcomb's problem. Therefore the player in each problem in outcome alignment should play the same way he would if the two problems were unconnected.

The standard Newcomb's problem NP is outcome alignable with CA. In the ordinary operation of NP the predictor's success or failure on any trial is independent of its successes and failures on other trials; Newcomb's problem thus generates two B(0.9) processes, one from its one-boxers, the other from its two-boxers. This is precisely what CA requires from its external source. On any given trial in which the predictor errs, the allocation is switched in CA. On trials in which the predictor scores a success, if both players one-box, both receive 1,000,000; if both players two-box, both receive 1,000. On trials in which the predictor errs, if both players one-box, they both receive 0; if both players two-box, they both receive 1,001,000. On a given trial, if the games are played the same way, then the outcomes are identical. (If the games are played differently, the outcomes are of course different.)

By the same reasoning NP(p, p) is outcome alignable with CA(p, p) for any p. If $p_1 \neq p_2$ the alignment construction is slightly different. The player choice in both games is always known before the box preparation in CA(p_1, p_2) is determined. If the players play the same way, proceed as above; if the players play differently, then consult some other B(p_1) or B(p_2) process. NP(p_1, p_2) is therefore outcome alignable with CA(p_1, p_2).

When NP and CA are in outcome alignment the following is true on every trial: if the advisor recommends one-boxing in NP, she knows her player will receive the same outcome as someone one-boxing in CA. If she recommends two-boxing in NP, she knows her player will receive the same outcome as someone two-boxing in CA. It is certain that one-boxing is best in CA. When in outcome alignment with CA, NP should be one-boxed, but in such alignment NP is no different that any other Newcomb's game. Therefore, against all orthodoxy, one-boxing is best in Newcomb's problem. The fools are correct.

To summarize, Newcomb's game is outcome alignable with CA. According to any decision theory, Newcomb's game alone should be played in the same way as when in this alignment. Since CA unquestionably should be one-boxed, so should N. This resolves Newcomb's problem the target paradox of this chapter. Readers uninterested in the solution to other cooperation problems can safely proceed to the next chapter.

5.5 Coherence

Outcome alignment can only occur if the problems have the same *possible* outcomes. We take the **standard outcomes** to be those of Newcomb's problem: $0 - 1,000 - 1,000,000 - 1,001,000.$[6]

Proposition 1: Any problem P with standard outcomes is outcome alignable with CA(p_1, p_2) for some p_1 and p_2.

Proof In P let p_1 be the probability of receiving 1,000,000, if C is played, and p_2 the probability of receiving 1,000 if D is played. If P is used to produce the Bernoulli processes B(p_1) and B(p_2) needed to source CA(p_1, p_2), then P and CA(p_1, p_2) will be in outcome alignment. On any trial if the players both choose C in their individual games, there is a probability p_1 both receive 1,000,000 and a probability $1 - p_1$ both receive 0; if both choose D, there is a probability p_2 both receive 1,000 and a probability $1 - p_2$ both receive 1,001,000.   □

---

[6] They are the clearest; the Solomon story is tarnished by the staggering immorality of Solomon's using his royal status to coerce a married woman into adultery, the prisoner's dilemma by gains obtained through betrayal, both of which the analyst is supposed to ignore because no evaluation is assigned to these moral matters in the problem's specifications.

C and D are the player choices; denote by C_0 or D_0 the cooperation or defection of the "other" which may be a player (Prisoner's Dilemma) an entity (Newcomb's problem) a player characteristic (charisma) a chance device, etc. If players or advisors are sampled with replacement from a fixed population, the associated choice-outcome pairs in the experimental data represent independent sampling from a stationary process, that is, the process does not change during the sampling. Essentially three numbers can be estimated from such data. r, the relative proportion of cooperators is a feature of the population not a feature of the problem. The other two, crucial to choice-outcome correlations, are $P(C_0|C)$ and $P(D_0|D)$ We call these probabilities the **advisor probabilities** p_1 and p_2 or the **player probabilities** q_1 and q_2 depending on whether advisor sequences or unadvised player sequences are at issue.

We have seen that problems in outcome alignment are for practical purposes the same. Decision theory is intended to give recommendations that are *practical* in the context of the problem. To agree with experimental results a decision theory must give the same solution to problems in outcome alignment.

The **Coherence Principle**: problems in outcome alignment should be played in the same way.

A decision theory is **coherent** if it assigns the same optima to outcome alignable problems.

Proposition 2: Causal decision theory, as well as the various reformed evidential theories, are incoherent.

Proof All of the aforementioned recommend one-boxing in CA and two-boxing in NP, which are outcome alignable.   □

Let v_1 through v_4 be the four possible outcomes, namely,

$$v_1 = v(C, D_0)$$

$$v_2 = v(D, D_0)$$

$$v_3 = v(C, C_0)$$

$$v_4 = v(D, C_0)$$

where v is value to the player. The expected values of the player choices are

$$E(C) = P(D_0|C)v_1 + P(C_0|C)v_3 = (1 - p_1)v_1 + p_1 v_3$$

$$E(D) = P(D_0|D)v_2 + P(C_0|D)v_4 = p_2 v_2 + (1 - p_2)v_4$$

Given v_1 through v_4, the question of optimality is completely determined by the values of p_1 and p_2. With the standard outcomes $E(C) > E(D)$ if and only if $p_1 + p_2 > 1.001$.

A problem is a **cooperation problem** if the player's choices C or D and the contribution of the "other" C_0 or D_0 are such that: (1) relative to a particular contribution of the other, D is better for the player than C, (2) relative to a particular choice of the player's, D_0 is worse for the player than C_0. Then $v_1 < v_2 < v_3 < v_4$. Unless noted otherwise all problems we consider are cooperation problems. (1) assures that the causal theory advises defection in all cooperation problems.

Proposition 3: In a coopertion problem if $p_1 + p_2 \leq 1$, then defection is optimal.

Proof

$$
\begin{aligned}
E(D) - E(C) &= p_2v_2 + (1 - p_2)v_4 - (1 - p_1)v_1 - p_1v_3 \\
&= p_2v_2 + v_4 - p_2v_4 - v_1 + p_1v_1 - p_1v_3 \\
&= v_4 - v_1 + p_1(v_1 - v_3) + p_2(v_2 - v_4)
\end{aligned}
$$

Since $p_1 + p_2 \leq 1$,

$p_1(v_1 - v_3) + p_2(v_2 - v_4) \geq (p_1 + p_2)\min(v_1 - v_3, v_2 - v_4) > (p_1 + p_2)(v_1 - v_4) \geq v_1 - v_4$

so $E(D) - E(C) \geq 0$. □

If $p_1 + p_2 > 1$, then the optimal course depends on the v_i's. For standard outcomes, cooperation is optimal if $p_1 + p_2 > 1.001$.

We can now solve the key problems:

(1) In the Solomon story the player probabilities $q_1 = P(\bar{R}|\bar{S})$ and $q_2 = P(R|S)$ are high since $\bar{R}$ and $\bar{S}$ are both caused by C, and R and S are both caused by $\bar{C}$. Use of advisors severs this correlation of summoning to revolution because revolution is governed by the player's charisma status, but the choice is directed by the advisor. In this case $p_1 = P(\bar{R}|\bar{S}) = P(\bar{R})$ and $p_2 = P(R|S) = P(R)$ so $p_1 + p_2 = 1$. Summoning (defection) is therefore optimal.

(2) In the tribal PD the player probabilities are both 0.9, and hence sum to 1.8. By this calculation cooperation is optimal since it gives a high probability to the other player's cooperation. However, an advisor for the designated player would sever the correlation between the designated player's choice and the other player's tribal affiliation. $p_1 = P(C_0|C) = P(C_0)$ and $p_2 = P(D_0|D) = P(D_0)$ so $p_1 + p_2 = 1$, and confession is optimal. Each player would like the other to cooperate, but there exist no means of bringing this about. Newcombian problems are different in that the player and predictor can each foster cooperation in the other, albeit by non-causal means.

(3) In Newcomb's problem the expected value calculation based on player sequences shows one-boxing to be superior. The predictor's task with advisors is the same as its task with players. Under the rules the player makes what he takes to be the best choice, and the advisor recommends what she takes to be the best choice. There are no grounds whatever for claiming that the predictor, so accurate with players, should be helpless against advisors. One-boxing, which is to say cooperation, is therefore optimal.

A coherent decision theory is one whose solutions are consistent with those derivable from advisor sequences. Since advisor sequences of sufficient length give a solution to every well defined problem, there is only one coherent decision theory, which we denote by **D**. R(**D**) consists of those act-outcome correlations generated by random advisor sequences. If R(**D**) $\not\subset$ R(**T**), **T** assigns different optima to a pair of outcome alignable problems; if R(**T**) $\not\subseteq$ R(**D**), **T** gives a recommendation based on correlations that do not appear in advisor sequences and hence do not pertain to recommendations. **D** is the only decision theory that avoids both pitfalls.

According to **D**, three kinds of correlation need to be distinguished: **causal**, **mimetic**, and **specious**. "Mimetic" refers to non-causal correlations that survive the transition from player to advisor sequences; "specious" refers to non-causal correlations that are eliminated by this transition. If the goal is to manipulate outcomes by means of decision, mimetic manipulation of outcome by act, whatever its philosophical status may be, is as reliable as causal manipulation.

5.6 The Perils of Incoherence

By **black box data** for a problem we mean choice-outcome data from an *advisor* sequence for the problem, in which only the choices, C or D, and the outcomes are displayed. Given unlimited amounts of black box data for a problem, what can be inferred about the optimal choice? According to any incoherent decision theory the surprising answer is: practically nothing.

Two problems are **distinguishable** if an observer can distinguish black box data from the problems on the basis of the appropriate statistical tests for Bernoulli variables (or simply because the possible outcomes are different). If the outcomes were manipulable to a different degree in two problems, this could be determined from black box data, that is, the problems would be distinguishable.

In this idealized setting even the slightest difference in the advisor probabilities of two problems would eventually be detectable from the problems' black box data. We can therefore define P with advisor probabilities p_1 and p_2 to be **indistinguishable** from P′ with advisor probabilities p_1' and p_2' if and only if $p_1 = p_1'$ and $p_2 = p_2'$ and the possible outcomes of the two problems are the same. If two problems are indistinguishable, then C *fosters* a given outcome in one problem to the same degree as it does in the other problem, similarly for D and some given outcome.

Although indistinguishability makes no reference to probability source arrangements or to correlation between the problems, indistinguishability is nevertheless closely related to outcome alignability. If P and P′ are indistinguishable, they have the same advisor probabilities p_1 and p_2, and are each outcome alignable with $K(p_1, p_2)$. Hence indistinguishability is the transitive closure of outcome alignability.

Suppose **T** is incoherent; then **T** rejects some mimetic correlation or accepts some specious correlation or both. Given only black box data, it is impossible to tell if it reflects mimetic correlations that **T** would reject or eliminates specious correlations that **T** would accept. This can be true for any black box data and can make a difference to optimality. The proponent of incoherent **T** must maintain that one cannot at all determine the optimal choice from unlimited amounts of anonymized data; one needs to look into the black box. Decision theoretic incoherence has a powerful anti-empiricist strain. Among decision theories only **D** is qualified to assess black box data.

Black box data may be insufficient to investigate the internal functioning of the process generating the act-outcome pairings, however such data tells the decision maker precisely what she needs to know, namely, how reliable a given action is for assuring a given outcome, when the action is performed under the guidance of a theory of optimality. The inability of incoherent decision theories to make principled use of black box data is a severe drawback.

Summary: Coherence is defined as agreement with the results of certain experiments that test recommendations. There is only one coherent decision theory. None of the traditional formulations are coherent. The coherent theory recommends taking one box in Newcomb's problem yet confessing in the prisoner's dilemma. An incoherent decision theory is incapable of drawing conclusions from unlimited amounts of anonymized act-outcome data.

References

Eells, E. (1982). *Rational Decision and Causality*. Cambridge: Cambridge University.
Eells, E. (1984). Metatickles and the dynamics of deliberation. *Theory and Decision*, *17*, 71–95.
Eells, E. (1991). *Probabilistic Causality*. Cambridge: Cambridge.
Gibbard, A., & Harper, W. (1978). Counterfactuals and two kinds of expected utility. Foundations and Applications of Decision Theory, Hooker, Leach & McClennen (eds.) vol. 1, D. Dordrecht: Reidel, 125–162. Reprinted in Gardenfors & Sahlin 1988, Decision, probability and utility and with abridgement in Cambell & Sowden 1985. Paradoxes of rationality and cooperation.
Jeffery, R. (1965). *The Logic of Decision*. New York: McGraw-Hill.
Jeffery, R. (1983). Revised edition of Jeffery (1965) University of Chicago, Chicago.
Kyburg, H. (1980). Acts and conditional probability. *Theory and decision* 12, 149–171.
Lewis, D. (1981). Causal decision theory. *Australasian Journal of Philosophy*, *59*(1), 5–30. Reprinted in Lewis; 1986, and in Gardenfors & Sahlin; (1988).
Skyrms, B. (1982). Causal decision theory. *The Journal of Philosophy, 79*, 695–711.

Chapter 6
The Open Box Game

The **Open Box game** (OB) is a standard Newcomb problem except that the 1st box is open; the player can see whether the 1st box is full or empty before making a boxing decision. There are four possible strategies in the Open Box game:

- The **Saint**: take one box no matter what you see in that 1st box, i.e., cooperate even in the face of the Predictor's defection.
- The **Good Sport**: take one box if it contains the million, take two boxes if the first box is empty, i.e., cooperate if the Predictor does.
- The **Spoiler**: take one box, if it is empty, take two boxes if the first box has the million, i.e., play so as to make the Predictor incorrect.
- The **Defector**: take two boxes no matter what.

Proposition 4: If a decision theory recommends the saint strategy in OB, then it recommends one-boxing in Newcomb's problem; if it recommends the defector strategy in OB, then it recommends two-boxing in Newcomb's problem.

Proof Newcomb's game is a probability mixture of the two cases in the open box game. By Savage's sure-thing principle, since the saint (respectively the defector) make the same choice for each component of the mixture, that same course is optimal for the mixture itself. Since the sure-thing principle is itself a dominance principle, we repeat the argument more carefully. With the sport or spoiler strategy, the player can declare the strategy in advance, but to reduce this strategy to a *choice* (take one or two boxes) the player has to see or otherwise know which of the two situations he faces. For the saint or defector strategy, the player makes the same choice in either situation, and can declare this in advance of seeing the open boxes. But these are exactly the circumstances of a player in Newcomb's problem. □

The open box game is not completely defined without specification of the predictor's intended responses to the four strategies. This is best discussed in terms of predictor motives. In Newcomb's problem the predictor's motive can be to maintain its success rate or to induce the player to one-box (for which it needs to maintain a good success rate). These lead to the same conclusions, which is not the case when these predictor motivations are transferred to the open box game.

W. Eckhardt, *Paradoxes in Probability Theory*, SpringerBriefs in Philosophy,
DOI: 10.1007/978-94-007-5140-8_6, © The Author(s) 2013

(1) The predictor's goal is accurate prediction. Predicted saints would receive the million, predicted defectors the empty box. Since the predictor cannot fail against a sport nor succeed against a spoiler, their payouts are indeterminate; if the predictor is presumed to refrain from giving out prizes for no reason, then predicted sports or spoilers receive an empty box. This interpretation has two defects: (a) The predictor cannot maintain a success rate of p if the fraction of spoilers in the player population is greater than $1-p$. (b) In the open box game as distinct from NP or CA the predictor's box preparation has a causal influence on the player. There is no monetary incentive to be a spoiler, and many would-be saints, when confronted with an empty box, would defect. With a normal human population the predictor should be able to exceed a success rate of 90 % simply by giving everyone an empty box. Ironically the emphasis on the predictor's success rate trivializes it.

(2) The predictor's goal is to induce players to one-box. Predicted sports get the million; predicted defectors and spoilers the empty box. The payout for saints is indeterminate. The predictor can afford to give a true saint an empty box; however, if a predicted saint is more likely to lapse on an empty box, the predictor might elect to give a predicted saint the million.

We adopt the second predictor motivation for a predictor that is 90 % correct with one-boxers, and more successful with two-boxers (as explained above, this is easy). If the 1st box contains a million, the player probabilities are $p_1 = 1$ and $p_2 = 0$; if the 1st box is empty, the player probabilities are $p_1 = 0$ and $p_2 = 1$. In either case they sum to 1 indicating defection which is two-boxing. To determine the advisor probabilities consider that the advisor is not influenced by the player with whom she is randomly paired; the advisor must accordingly be prepared to give advice for either case, a full or empty 1st box—she has to have settled on one of the four strategies. There is no monetary rationale for recommending the spoiler strategy; an advisor who used the player to spite the predictor would break the rules of advisorship. Players who are advised to be sports (and are hence bound to that strategy) fare decidedly better than those advised to be defectors and have more assurance than those advised to be saints. It comes down to sport vs. defector (perhaps with the stray saint). Advisor probabilities are $p_2 > p_1 = 0.9$, that is, $p_1 + p_2 > 1.8$, indicating cooperation—one-boxing on a full 1st box.

In OB there is a more than the usual split between intention and action. In the other problems there is no reason to expect a player to act otherwise than he intended. In OB many players who sincerely intend to be sports would upon seeing the million be overcome by the idea that they could now take the 1,000 with impunity. Such players are likely to receive an empty 1st box in the first place. Effective play in OB requires a fusion of preliminary intention and subsequent action. There are telling parallels with story of Odysseus and the Sirens. Odysseus can safely listen to the Sirens' song only if he can prevent his own mid-course defection, otherwise listening will be disastrous. Only by binding himself to his original strategy, much as his crew binds him to the mast, does he succeed. Another example: suppose two medical treatments administered 24 h apart are effective in curing a certain illness, but the first treatment alone has negative side effects, including development of a phobia against taking the second treatment. The second treatment neutralizes these side effects, and cures the illness. Taking the first treatment is foolhardy without making provisions to have the second forced upon you. Like Odysseus and like the player in OB, it is crucial to prevent your own defection.

The coherent and the evidential theorist agree that the sport strategy has the greatest expected value. However, once the open boxes are presented, the evidential approach succumbs to a Siren song: no matter what may have originally been optimal, it is now better to take two boxes. Conditional on current circumstances, the expected value of two boxes is indeed a thousand more than the expected value of one box. The evidential theorist undergoes a mid game change of perspective.[1] In advance of setting up the game the evidential theory assesses the sport strategy to be the best, but when the boxes are presented, it recommends defection. This falls short of logical contradiction, but such dithering cannot be sound. There is no rationale for planning one action, then performing another; in particular nothing in the problem's specifications suggests those who plan to be sports and then defect fare better than those who plan to be defectors and then defect.

Unless explicitly assigned a negative value, the psychological difficulty of making a certain decision is not supposed to detract from the value of that choice. If a player has a psychological compulsion to make the wrong choice, the counsel of decision theory is nevertheless to make the right choice. If player psychology interferes with analysis, it is the task of decision theory to circumvent this obstacle. The reasoning underlying the two-boxing defector strategy seems compelling because it is founded on a psychological compulsion that is so powerful it blinds two-boxers to the plain fact that one-boxers secure substantially better outcomes not only according to disputable expected value calculations, but by post game tallies of results. It may be worthwhile to tabulate a representative 60 trial sequence (see table). With a million in the 1st box, two-boxers do better; with an empty 1st box, two-boxers do better, and together these comprise all cases. Nevertheless on average one-boxers make 900,000 and two-boxers 21,000.

1st box	Choice	Outcome	No. of cases
Million	1 Box	1,000,000	9
Million	2 Box	1,001,000	1
Empty	1 Box	0	1
Empty	2 Box	1,000	49

---

[1] This should not be confused with the reasonable policy of updating a strategy in light of new information. Seeing the boxes provides no information that was not already taken into account in the original strategy.

In this problem dominance reasoning is so compelling that it is extremely difficult for the player, as distinct from the advisor, to see his way around it. The best way to play is for the player to make a binding contract in a public manner to follow the sport strategy or forfeit all proceeds. If this is done before the box preparation, it virtually guarantees the million; if it is done after the box preparation, it still nets the million nine times out of ten.

Chapter 7
The Hadron Collider Card Experiment

In a series of papers (Nielsen and Froggatt 1996; Nielsen and Ninomiya 2006, 2007, 2008, 2009) Holger B. Nielson and Masao Ninomiya present a reformulation of physical theory[1] which if correct would have paradoxical consequences for the long awaited production of a Higgs boson. According to this theory any attempt to produce a Higgs boson—a prime goal of the recently built Large Hadron Collider—is subject to anomalous disruptions from the future that are directed toward preventing this production.[2] Among these retrocausal influences (by which we mean influences in which the effect precedes the cause) the authors place the recent accident at the Large Hadron Collider, the political cancellation of the Supercollider by the United States Congress in 1993, and possible bankruptcies, political coups, and natural disasters, all of which have served or would serve to undermine production of this elusive boson. Since the latter could well be costly in terms of lives and property, the authors propose an astonishing **Collider Card Experiment**. Simulate a card drawing with one message card which we call the **jackpot card** and somewhere between 200,000 and 1 million blank losing cards. The jackpot card says "Stop the Collider".[3] CERN and other parties responsible are supposed to agree to stop all attempts at Higgs boson production if the jackpot card is drawn, the rationale being that this statistical near impossibility can only be interpreted as a message from the boson avoiding future. The authors envision this future constraint acting upon current attempts at Higgs boson production through a kind of path of least resistance, for example, the authors wish not to set the jackpot

---

[1] The Lagrangian possesses an imaginary part as well as the customary real part. The addition makes little difference (it tends to be self-cancelling) except near the Big Bang or in the vicinity of Higgs boson production. The value of this Lagrangian "pre-arranges" the initial conditions of the Universe for complete avoidance or severe restriction of this production.
[2] The future constraint does not absolutely preclude Higgs boson production but severely limits it. The argumentation in this article can be recast in terms of avoiding production of all but a very few Higgs bosons, replacing "no-boson" by "few-bosons", etc.
[3] Refinements to the card experiment have been proposed which give the future constraint other more nuanced messages to send. Since the arrangement described above reflects the essential feature of the experiment—the retrocausal message—we disregard these further complications.

W. Eckhardt, *Paradoxes in Probability Theory*, SpringerBriefs in Philosophy,
DOI: 10.1007/978-94-007-5140-8_7, © The Author(s) 2013

probability too low for the reason that if it were lower than the probability of a disaster that would destroy the Collider, the boson-avoiding future might favor the disaster over the card game as a means of preventing Higgs boson production. We refer to this radical formulation of physical theory or to some close analogue along with the perplexing conclusions the authors draw about Collider operation as the **Higgs anomaly**. In the likely event that the Higgs anomaly is non-existent, the experiment is a pointless exercise. Accordingly in preliminary discussion of the card experiment we presuppose the reality of the Higgs anomaly.

The authors candidly admit that their startling conclusions have a low probability of being correct; they even factor this into a cost-benefit analysis of the Collider Card Experiment. In the context of decision theory, what is most important is not the reality of the Higgs anomaly but the decision problem presented by it and the correct solution.

7.1 Another Open Box Game

The Card Experiment has the same structure as other problems considered: the player has a binary choice of stopping or not stopping the Collider; there are four possible outcomes: either a blank card or the jackpot card is drawn, either the Collider is stopped or not. If J represents drawing the jackpot card and H continuing to operate the Collider, the four outcomes in order of ascending value are: $\overline{JH}/JH/J\overline{H}/JH$. The Collider Card Experiment is a cooperation game, furthermore it is decision theoretically equivalent to the Open Box game.

The predictor operates from the past whereas the no-boson constraint operates from the future. However, the infallible predictor's uncanny ability to anticipate player choice means the predictor's earlier preparation of the box is determined by the player's later decision, thereby creating a mimetic influence of the Player's decision on the Predictor's earlier preparation. In decision theory one cannot make a distinction between how to behave in the face of a true retro-influence and how to behave in the presence of something that is rigged to act exactly as though there were a retro-influence. What matters is that both the Predictor and the future constraint can in a mysterious manner reward co-operation and punish defection.

The no-boson constraint acting from the future, is best modeled as 100 % successful like the infallible predictor. Appealing to the open box game further cements the analogy with the Card Experiment since the player knows whether he has drawn the jackpot card before making the decision whether to continue operating the Collider. What we can learn from OB is that knowing whether you've won in the Newcomb game before you make your boxing choice, makes it quite difficult to make the right choice. In the high stakes Collider Card Experiment, the temptation to defect is even greater: if you draw the jackpot card,[4] you would want to learn more by running the Collider anyway; you might also want to defect in order to foil the "infallible" future constraint. There would additionally be economic, institutional, and political incentives to continue running the Collider. The greater the benefits of continuing to run the Collider, the less credible is a decision to cooperate, and hence the less likely it is for the message to be sent.

The Collider Card Experiment is then an open box game with an infallible predictor. If a blank card is drawn, the 1st box is empty, otherwise the box contains the Jackpot card; the 2nd box contains the possibility of continuing to run the Collider. As always the player can take either the first box or both boxes.

The causal theory dictates that defection in the Card Experiment is the rational course. The Card Experiment attempts to trade the running of the Collider for the message from the future, but once the message from the future is received, one is free to continue running the Collider anyway. The alternatives are "having the message" and "having the message and continuing to run the Collider". The second is always worth more than the first, so defection after receiving the message is the optimal strategy. Continuing to run the Collider is the causally dominant choice no matter what card is drawn. The only way the causal decision theorist can receive the jackpot is randomly, something everyone agrees will not happen, and would be a meaningless fluke if it did, so the exercise is pointless. This inability to place any value in the card game should not be interpreted as a prescient rejection of the Higgs anomaly; decision theory is not equipped to make that judgement. Even if the Higgs anomaly exists and operates as the authors imagine, causal decision theory would demand the card game be conducted in a futile manner. Causal decision theory gives no useful guidance in the matter.

In the Card Experiment, the defector assures a blank card; the saint and the spoiler stop the Collider on a blank card, the worst idea of all. The optimal strategy is the sport—stop the Collider on the jackpot card, otherwise let the Collider continue to operate. This is the only strategy that has a shot at hitting the jackpot.

To clarify the underlying parallelism between the Collider Card Experiment and OB we group corresponding elements from each problem.

A. The future constraint corresponds to the Newcombian Predictor; these are what make a Newcomb's game possible.
B. The card drawing corresponds to the Predictor's box preparation—the jackpot card to a million dollar box, a blank card to an empty box—these are under the control of (A).

---

[4] It is only under assumption of the Higgs anomaly that the card game is a Newcomb game. In practice, drawing a blank card (empty first box) would be taken to disconfirm the Higgs anomaly and indicate that we were not playing a Newcomb game after all. In this case it would be folly to "cooperate". This makes the Card Experiment even trickier than the Open Box game since with the latter, one is sure at least to be playing a Newcomb game.

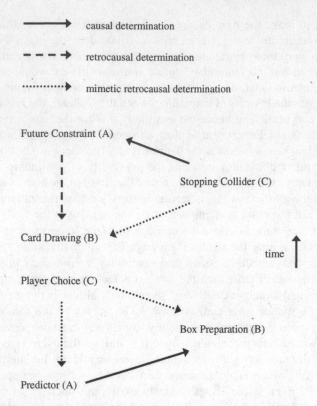

Fig. 7.1 (A) refers to Newcombian factors, (B) to matters under the control of (A), and (C) to matters under Player control

C. Stopping/not stopping the Collider corresponds to the Player's boxing choice; these are under the control of the Player.

Using arrows to represent causal or simulated connections, we see that in terms of the directions of the arrows [see Fig. 7.1] the diagrams for the two problems are the same: two arrows leave (C) and two arrows converge on (B). In both diagrams the (C) to (A) and (A) to (B) arrows *induce* a relationship between (C) and (B). The nature of this connection is determined by the arrows that induce it. This is the key both to Newcomb's problem and the card experiment. In both games the rest of the setup serves to induce a mimetic retrocausal relation between (C) and (B): it is *as though* one-boxing results in a previously filled box; it is *as though* stopping the Collider results in the earlier drawing of the jackpot card (Fig. 7.2).

Player/Other	Outcome	Newc.	PD	Sol.	Card
C D_0	v_1	0	10 years	R	0
D D_0	v_2	1,000	5 years	SR	H
C C_0	v_3	1,000,000	1 month	0	J
D C_0	v_4	1,001,000	0	S	JH

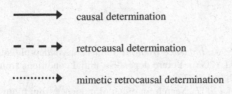

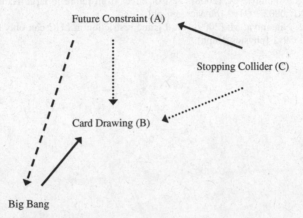

Big Bang

Fig. 7.2 According to the physical theory behind the Higgs anomaly, the (*A*) to (*B*) connection is itself *mimetic retrocausal*, induced by means of true *retrocausal* action of (*A*) upon the initial conditions of the Universe. Note that if the two mimetic arrows are removed, there remains a *causal* progression from stopping the Collider to drawing the jackpot card

For the player, cooperation (C) is one-boxing/not confessing/not summoning/ stopping the Collider; defection (D) is two-boxing/confessing/summoning/running the Collider. For the other, cooperation is full box/not confessing/charisma/jackpot card; defection is empty box/confessing/no charisma/blank card. The other symbols are: revolt (R), summoning (S), running the (Hadron) Collider (H), jackpot card (J).

	Decision theories	Evidential	Coherent	Causal
Problems	Tribal PD	C	D	D
	Solomon	C	D	D
	Newcomb's problem	C	C	D
	Open box game	C then D	C	D
	Collider card Experiment	C then D	C	D

References

Nielsen, H.B., Froggatt, C. (1996). *Influence from the Future*, arXiv:hep-ph/9607375v1.

Nielsen, H. B., & Ninomiya, M. (2006). Future dependent initial conditions from imaginary part in lagrangian, arXiv:hep-ph/0612032v2.

Nielsen, H. B., & Ninomiya, M. (2007). Search for effect of influence from future in large Hadron Collider, arXiv:0707.1919v3 [physics.gen-ph].

Nielsen, H. B., & Ninomiya, M. (2008). Test of effect from future in large Hadron Collider, a proposal, arXiv:0802.2991v2 [physics.gen-ph].

Nielsen, H. B., & Ninomiya, M. (2009). Card game restriction in LHC can only be successful!, arXiv:0910.0359v1 [physics.genph].

Part III
Mystifying Envelopes

Chapter 8
The Two-Envelopes Problem

The Two-Envelopes problem involves no questions of human nature or survival, nor any dubious assumptions such as human randomness. There are no issues of cooperation or defection, no dominance principles at stake. The problem is game theoretically trivial since the player has no opponent. It is a purely causal problem, so from the viewpoint of decision theory there should be no disagreements about how to play. Yet there is a tremendous amount of disagreement on this subject: at least three mutually inconsistent *false* approaches have been developed, as well as some minor misses. The Two-Envelopes problem which shouldn't be a paradox at all has nevertheless been the subject of arcane and perplexing disputes.

Unlike the other paradoxes, the Two-Envelopes game brushes against infinity, but all that is needed for its solution are a few facts about infinite series, known since the 19th century. Evidently the paradoxical implications of divergent series, long resolved in mathematics, retain the ability to bewilder in a probability context.

The blind game: You're presented two envelopes, identical in appearance, and are informed that one envelope contains twice as much money as the other. (To conceal quantities, these can be checks.) You are randomly allotted one (we call this the *1st envelope*) and then are offered the opportunity to switch envelopes. Since you have exactly the same knowledge concerning each envelope, the only reasonable conclusion is that you should value the envelopes equally, and hence be indifferent between switching or not switching. Alongside this is placed a paradoxical argument: suppose the 1st envelope contains F, then the other contains 2F half the time and F/2 half the time. So the other envelope is worth

$$(1/2)2F + (1/2)F/2 = 5F/4$$

an amount greater than F. This remarkable equation promises a 25 % average profit just from switching envelopes. However this line of reasoning collides with itself: we can as easily denote the amount in the 2nd envelope by F, then symmetrical reasoning yields

W. Eckhardt, *Paradoxes in Probability Theory*, SpringerBriefs in Philosophy, DOI: 10.1007/978-94-007-5140-8_8, © The Author(s) 2013

$$(1/2)2F + (1/2)F/2 = 5F/4$$

for the value of the 1st envelope, so by this account each envelope is worth 25 % more than the other.

The blind version of the paradox rests on a simple equivocation. The first occurrence of F refers to the case when F is the smaller amount, the second occurrence to the case when F is the larger amount. In a polynomial expression a variable must represent the same thing, whether known or unknown, at each of its occurrences. The symbol F has been used for two different random variables so the resultant expression is incoherent. Let S and 2S represent the unknown amounts in the two envelopes. The player's envelope is equally likely to contain S or 2S so the expected value of this envelope is 3S/2. Now it's true the other envelope is equally likely to contain half or twice the player's envelope: with probability 1/2 the other envelope contains half as much, but that is when the player's envelope contains 2S; this gives $1/2(1/2)2S = S/2$; with probability 1/2 the other envelope contains twice as much, but this is when the player's envelope contains S, giving $(1/2)2S = S$; these two sum to 3S/2. Both envelopes have exactly the same expected value. There is no expected gain in switching.[1]

The correct conclusion in the blind version can be derived from a fundamental principle, without appeal to expected values. Say an agent is **symmetrically ignorant** with respect to the choice of two options, if everything the agent knows about either option applies equally to both of them. It automatically satisfies this definition if the agent knows nothing about the options; on the other hand the agent can know quite a lot about how the options are similar and nonetheless be symmetrically ignorant as to some distinguishing feature. The **symmetrical ignorance principle** states that **symmetrical ignorance forestalls rational preference**. A warranted preference implies an asymmetry, surely a sound principle for rational decision making. As compelling as the expected-value argument may be, the proof based on the symmetrical ignorance principle is more fundamental and decidedly more general.

8.1 Opening the Envelope

The blind version, based on a miscalculation, stands as a cautionary example in the misuse of elementary expected value formulas, but the informed version, hinging on the subtleties of operating with the infinite, can be put to service in shattering some compelling illusions that arise when expected valued expressions form divergent series.

---

[1] The favorability of switching randomized envelopes is subjected to criticism in Bruss (1996); Bruss and Ruschendorf (2000), Clark and Shackel (2000) and Katz and Olin (2007) but the sophistication of the techniques and the complexity of the arguments obscure the simplicity of the fallacy.

The informed game: suppose the player is permitted to look into the 1st envelope before making the decision whether to switch. The amount in the 2nd envelope remains concealed until after the player's decision. Observing the amount in the 1st envelope breaks the symmetry of the game. Once the player learns F—the amount in the 1st envelope—she knows the 2nd envelope contains either F/2 or 2F. If, furthermore, she knows the probability distribution S from which the amounts were selected, she can determine the probabilities p and 1-p of these alternatives. The expected value of the other envelope is

$$p(F/2) + (1 - p)(2F)$$

This is greater than F if an only if (1-p) > p/2; that is, *it is favorable to switch when the probability of doubling is more than 1/2 the probability of halving*. One can find a distribution for which this relation holds for each value of F. For such distributions it would seem a winning policy to switch *irrespective of the value of F*. This policy can be fully implemented without opening the 1st envelope which brings us back to the first Two-Envelopes paradox and its insane advocacy for switching under symmetrical conditions.

Notation: i, j, k, n and N are nonnegative integers. I choose a single series of possible envelope contents $\{2^i\}_{i \geq 0}$. In other words the contents of an envelope is always a power of 2, including $2^0 = 1$. Let S_i denote that 2^i is *selected* hence 2^i and 2^{i+1} are placed in the envelopes, F_j that 2^j is *found* in the 1st envelope; $S_i F_j$ means they happen together. Since S_i implies either F_i or F_{i+1}, $P(S_i F_j)$ is nonzero only if j = i or j = i + 1. SW refers to the strategy of always switching, PS to the strategy of always passing (= not switching). Let $P_i = P(S_i)$ be the probability 2^i is drawn. To make it favorable to switch in each case, we need to assure $P_{i+1} > P_i/2$. The easiest way to accomplish this is to keep the ratio P_{i+1}/P_i constant by giving the S_i amounts a geometric distribution $P_i = (1-r)r^i$ with 1/2 < r < 1. For numerical results I use r = .75. Both switching and passing net more money than not playing, hence the profitability of switching in any context should be judged relative to that of passing. A strategy X should be assessed relative to the benchmark performance of PS; it's the difference X-PS that matters. By the favorability or profitability of a strategy X we always mean the favorability or profitability of X-PS. The player knows the distribution from which the envelope amounts have been selected and can calculate the relevant probabilities.

I've chosen to prove results for an infinite class of geometric distributions with special attention to the case r = .75. This makes arguments less abstract. This reasoning can be generalized to any distribution for which it is individually favorable to switch on a chosen amount, no matter what that amount is. These include all the distributions for which there arises a Two-Envelopes paradox. In other words this treatment is adaptable to any Two-Envelopes paradox.

Consider two expected value computations: (i) For any S_i selected the 1st envelope contains 2^i half the time and 2^{i+1} half the time. Then SW − PS makes 2^i half the time (when $F_i = 2^i$) and loses 2^i half the time (when $F_i = 2^{i+1}$). For *any* S_i switching and passing are worth exactly the same: $E(SW -PS|S_i) = 0$. (ii) When the 1st envelope contains $F_0 = 1$, switching is a sure profit. For $j \geq 1$, the expected value of switching on F_j is $2^{j-1}(2r-1)/(1 + r)$ (see Sect. 9.6). For r = 0.75 this is $(0.29)2^{j-1}$, which is positive and increases exponentially in j. From (i) we learn that SW and PS are equally good for all S_i; from (ii) that SW is better than PS for all F_j. Yet the S_i and F_j series each cover all possible cases, albeit with different groupings. It's worthwhile to compare these conflicting expected value calculations to those of Newcomb's problem. In Newcomb's problems the conflicting calculations proceed under different rules—one incorporates mimetic correlations and the other ignores them, so of course they come to different results. In the Two Envelopes the collision is head-on. The calculations are of the same kind taking into consideration the same factors. Moreover, (i) and (ii) are completely routine, in fact unassailable, yet they appear to come to contradictory conclusions. This is the (informed) Two-Envelopes paradox which has prompted a number of inequivalent, indeed mutually inconsistent, resolutions. In order to sort through this material I identify three pivotal assertions; all major approaches to the paradox are determined by the truth values assigned to these assertions.

EC (each case): *in the open game it's favorable to switch on each F_j considered individually*. This is because $E(SW - PS|F_j) > 0$ for each F_j. The validity of **EC** is assured by the choice of distribution, virtually by stipulation. **EC** like each of the other two assertions, has been rejected by some parties to the debate, but it has never been explained what could possibly be wrong with it. There is an occasional dark pronouncement about the perils of infinite expected value, but once the player learns the amount in the 1st envelope, the ocean of unrealized possibilities of what might have been in the envelope is removed from an expected value assessment. Infinite expected values evaporate once the 1st envelope is opened, leaving **EC** free of the infinities and divergences that can complicate analysis of **SYM** and **AL**. The argument that it's profitable to switch on an *individual* F_j is routine, finitary, and airtight.

SYM: *in the informed game the strategies of switching for all F_j or of passing for all F_j each finish with the larger amount half the time and the smaller amount half the time; SW has absolutely no advantage over PS*. The following reasoning establishes **SYM**. The symmetrical ignorance principle shows that SW has no expected advantage over PS in the blind game. The informed-game player is not symmetrically ignorant, but by playing an unconditional strategy, SW or PS, she achieves exactly the same results as her blind-game counterparts. The randomization of the envelopes deprives the player of reasons to prefer SW or PS. Since SW and PS have infinite expected values, some care must be taken in comparing them; **SYM** asserts that SW and PS are *equally good* in the sense that they have precisely the same distribution of outcomes—the same outcomes with the same frequencies.

AL (always): *in the informed game the strategy of always switching, no matter what amount is found in the 1st envelope, is a better strategy than passing—SW is better than PS*. **AL** states that SW is better than PS, whereas **SYM** claims they are perfectly equivalent; thus **AL** contradicts **SYM**. The inflexible SW strategy can be carried out without opening the envelope, so **AL** justifies the profitability of switching in the blind game. Proponents of **AL** face the hopeless task of avoiding this unwelcome conclusion (see the 1st and 3rd approaches below) (Fig. 8.1).

Approach	Accepts // Rejects	Drawback	Recommended Strategy
1st	EC, SYM, AL	self-contradictory	anything
2nd	SYM // EC, AL	violates expected value principle	never switch
3rd	EC, AL // SYM	violates symmetric ignorance principle	always switch
4th	EC, SYM // AL	——	$M(2^n)$

Fig. 8.1 Summary of approaches to the paradox

We have seen that **AL** and **SYM** contradict one another. The other crucial relationship is that between **EC** and **AL**, specifically does **EC** imply **AL**? In briefest terms the argument for **EC** → **AL** is: if it is good to switch in *each case*, then it is good *always* to switch. Logical interrelationships among these statements are delineated in the following.

Proposition 1: Among **EC**, **SYM**, and **EC** → **AL** the conjunction of any two implies the third is false.

Proof: (i) Given **EC** and **EC** → **AL**, **AL** follows which contradicts **SYM**. (ii) Given **SYM** and **EC** → **AL**, **SYM** implies ¬**AL**; the conjunction of **EC** → **AL** and ¬**AL** implies ¬**EC**. (iii) Given **EC** and **SYM**, **SYM** implies ¬**AL**; the conjunction of **EC** and ¬**AL** contradicts **EC** → **AL**.   □

The path to the correct solution has been cleared. **EC** and **SYM** are well established through independent arguments (expected value calculations and the symmetric ignorance principle, respectively) while **EC** → **AL** places a wedge of contradiction between them. Moreover the only reason to believe **AL** is the combination of **EC** and **EC** → **AL**. Therefore, **EC** and **SYM** are true; **AL** and **EC** → **AL** are false. But strong conceptual forces obscure this path; commentators have made mystifying, preposterous or self-contradictory claims about the problem. Before examining the consequences of this solution, we review the history of the paradox in terms of three mistaken approaches, distinct attempts to avert or suppress the inconsistency and absurdity which follow upon acceptance of **EC** → **AL**.

1st Approach: All three assertions are true. Historical recommendation: switching in the blind game yields no advantage, but always switching after the envelope is opened does yield an advantage.

This attempt to retain all three assertions is hopelessly self-contradictory[2] which may explain why its proponents tend to express dissatisfaction with it. For instance

---

[2] It's not unprecedented to hold contradictory conclusions when it comes to the infinite. Compare three Medieval reactions to the paradox that two infinite magnitudes are unequal if one

Sobel (1994, p. 94) claims there are conditions under which you should always prefer to pass before opening the envelope and always prefer to switch after opening it but concedes this "makes you less than a perfect practical intellect". Brams and Kilgour (1996) hold that symmetry makes always switching and always passing equivalent yet that under certain conditions it's always favorable to switch; they say this "cries out for further explanation". Efforts at reconciling the contradictory belong to the early history of the problem; 21st century commentators have settled for other absurdities.

2nd Approach: **SYM** is true, but **EC** and **AL** are false. Recommendation: whatever amount appears in the 1st envelope, switching and passing are equally good.

If **SYM** and **EC** → **AL** are accepted, then since **AL** contradicts **SYM**, **EC** must be rejected. In this supposed contest between **EC** and **SYM** the intuitively transparent **SYM**, requiring no calculations for its justification, appears to cut through all the knotty details and demonstrate the equivalence of switching and passing once and for all. Proponents of this approach maintain that **EC** rests on some kind of illusion. Clark and Schackel (2003, pp. 699–700) write "opening the envelope makes no difference to the contents of the envelopes, so cannot make any difference to the correct application of decision theoretic calculations." This is sheer nonsense. Suppose a card is dealt face down to each of two players, to be followed by a round of betting on who has the higher card. Apply the aforementioned rule to this case: looking at my card makes no difference to my card or my opponent's card, so it cannot make any difference to the correct application of decision theory. Observing the amount in the envelope introduces the possibility of selective strategies that are better than SW or PS and hence improve the expected value of subsequent play. The authors themselves give evidence against their astonishing futility claim: "If all those cases where you have 2 in your envelope are picked out then the average gain for them is likely to be positive" (Clark and Schackel 2000, p. 430). This essentially concedes **EC** which nudges their position into internal contradiction. To defuse the admission they continue "In considering the average gain for a given value in your envelope we are not considering a truly representative sample, one for which we are as likely to have the larger sum in our envelope as the smaller." A sample consisting only of instances of switching on 2 is indeed not representative of general switching; that's what allows the case of switching on 2 to be profitable. In other words, by looking into the 1st envelope the player can segregate those non representative cases of F = 2 for special treatment. This secures the positive average gain the authors concede for switching on 2 and otherwise gains zero by passing (see 8.7.1). Switching only on 2, which requires looking into the 1st envelope, beats any strategy that can be implemented without knowledge of F. Precisely the same reasoning applies to the strategy of switching

---

(Footnote 2 continued)
is a proper part of the other, yet equal because both are infinite: (1) infinites do not exist (2) "equal" and "unequal" do not apply to infinites (3) infinities can be at once equal and unequal (Kretzmann et al. 1982 pp. 569–571).

only on a particular higher power of 2. Each of the F_j represents a subsample that is nonrepresentative of switching generally.

3rd Approach: **EC** and **AL** are true, but **SYM** is false. Recommendation: always switching is better than always passing.

If **EC** and **EC** $\rightarrow$ **AL** are accepted, **AL** follows, and **SYM** must be rejected. This skirts the internal contradictions of the 1st approach; the cost is a preposterous violation of the logic of symmetric ignorance. Meacham and Weisberg (2003) adopt aspects of this approach but strive to evade its unpalatable consequences. In an attempt to avert the clash between **SYM** and **AL**, they assign each a separate area of application. Symmetry considerations are claimed to apply only to the blind game whereas unconditional switching is advocated only for the informed game.

The most direct way to disprove **AL** is as follows: **AL** says it is favorable always to switch in the open game. Since this inflexible strategy can be carried out without opening the 1st envelope, it is favorable to switch in the closed game also, but this contradicts the symmetrical ignorance principle. The authors seek to forestall a *reducio ad absurdam* by denying that **AL** carries this implication for the blind game; specifically they claim it is a profitable strategy to switch in the informed game but not profitable to do the same in the blind game. They defend this perplexing combination as free from "inconsistency" or "logical contradiction". The matter can be resolved by going beyond mere logical consistency to include everyday causal realism. We are accustomed to the idea that additional information, such as the amount in the 1st envelope, can improve one's results in a game, but the authors claim the information that allegedly brings about the improvement need not be used in any way. This faces the unanswerable objection that opening the envelope or learning its content would need to exert a miraculous influence over the outcome of switching. Unconditional switching has to work equally well in the blind game as in the informed game. By the symmetrical ignorance principle it gives no advantage in the blind game, so **AL** is false. The affirmation of **AL** in this approach is a mistake from which there is no recovery.

For Clark and Shackel learning the amount in the 1st envelope is of no use, hence of no benefit to the player. For Meacham and Weisberg as well as Dietrich and List (see Sect. 9.6.1) learning the amount benefits the player but in an unexplained manner that does not entail use of this information. In the next approach it is shown that learning the amount in the 1st envelope is beneficial if this information is used to improve play.

4th Approach: As intimated above, **EC** and **SYM** are true, but **AL** and **EC** $\rightarrow$ **AL** are false. Recommendation: SW and PS are perfectly equivalent, but there are better strategies.

The strategy SW $-$ PS has no expected value. The series corresponding to the expression $E(SW - PS)$ is divergent in the sense that the sum fails to exist (in contrast, the expected value of SW or PS diverges to an infinite positive sum.) It's been known since the 19th century that grouping the terms of a divergent series can create a convergent series and that different ways of grouping the terms can lead to different sums. Let $\sum_{i,j} (P_i/2)\big(E(SW|S_iF_j) - E(PS|S_iF_j)\big)$ be a series

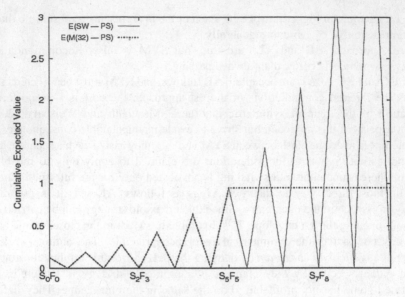

Fig. 8.2 E(SW − PS) vs. E(M(32) − PS)

expansion of E(SW − PS). Suppose terms with the same S_i are grouped. Since SW and PS have exactly the same expected values on 2^i and 2^{i+1} each gaining half the time and losing half the time, the sum of the terms in each S_i group is zero, so all groups sum to zero. If terms with the same F_j are grouped, the result is starkly different. The active strategy of playing for 2^{j+1} fares better than the passive one of keeping 2^j. Each group sums to a higher positive number than the one before it; the sum for all these groups is infinite. These calculations involve no trickery or distortion: the infinite sum F-grouping points to the truth of **EC** just as the zero sum S-grouping points to the truth of **SYM**; there is no inconsistency since the original series has no sum.

Clark and Schackel (2000, p. 426) write "In the paradoxical envelope cases only one of the three results on swapping—(1) positive average gain, (2) zero average gain, (3) average loss—can be correct". There is a fourth answer—there is no *average* gain hence it's neither positive, negative, nor zero. The series corresponding to E(SW−PS) is divergent since only divergent series can be bracketed in ways that produce different sums.

In Fig. 8.2 the solid line is a graph of the cumulative expected value of SW−PS. Summing conditional on F-values catches this sequence at its peaks, summing conditional on S-values catches the sequence at its troughs. The advantage of switching on 2^j and doubling is canceled by the disadvantage of switching on 2^{j+1} and halving, since both events have probability $P_j/2$. Although switching on an individual F_j truly has positive expected value, switching for every F_j cancels all advantages. An infinite number of advantageous opportunities can indeed "add up" to breaking even.

As paradoxical as a disparity between *better for each case* (**EC**) and *better overall* (**AL**) may seem, it originates in the fact that alternative groupings of the terms of a divergent series can have different convergence properties or different sums. Ordinary language is ill equipped to maintain the distinction, since with finitely many alternatives "better for each case" does indeed imply "better overall". An expression such as "it's favorable to switch for every amount" can shift between **EC** and **AL** like a semantic Necker cube.

EC → **AL** is based on a fallacy of composition: individually profitable strategies that do not interfere with one another can always be assembled into an overall profitable strategy. The informed game illustrates that on the contrary the composition of an infinite number of profitable selective strategies can yield an unprofitable, unselective strategy. It demonstrates the hold the paradox retains that a well defined problem with exact probabilities and payoffs is thought to demand, not standard probability theory, but original principles and new techniques. Scholars have displayed ingenuity in devising ways to defuse **EC**, circumvent **SYM**, or blunt the disruptive potential of **EC** → **AL**. This has served only to prolong the career of the paradox. This problem requires not innovation but application of long accepted principles. If a problem can be solved using the standard axioms of probability theory along with the theory of infinite series, it is gratuitous to resort to novel, exotic, or untried techniques. In particular the problem does not call for uniquely Bayesian analysis (Christensen and Utts 1992) Markov kernels and conditionally specified distributions (Bruss and Ruschendorf 2000) non standard utility functions, (Horgan 2000) nested expected value formulas and variant decision theories (Clark and Schackel 2000) or subjunctive conditionals defined over possible worlds (Katz and Olin 2007).

Summary: under symmetrical circumstances as in the blind game, switching has no advantage over passing. Learning the amount in the 1st envelope breaks this symmetry, and each possible envelope amount reveals an asymmetric opportunity for expected gain in switching. If the informed-game player elects to exploit *all* these opportunities by adopting a pure switching strategy, the asymmetries cancel one another, restoring symmetry and erasing the profitability of switching.

8.2 How to Play

Given the energies devoted to theoretical disputation on the two-envelopes game, it's odd that antagonists don't try playing the game. The erroneous approaches outlined all conclude there is no way to profit from knowledge of the specific amount in the 1st envelope.[3] Whatever reasons are given for switching on a particular 2^j seem to transfer to all other 2^k, leaving the impression that the only

---

[3] A similar tendency to disregard useful information is seen in the more elementary Monte Hall problem. One of three boxes contains a (fixed) prize and the player is randomly given one box. A moderator who knows the location of the prize reveals one of the other two boxes to be empty; it's always possible to do this. The player is offered the choice to switch for the remaining unopened box. One time in three she already possesses the prize and loses it by switching but two times out of three, she has an empty box and wins the prize by switching. Nonetheless many who

principled strategies are the inflexible ones—always switching or always passing. To see this is not the case we consider multiple rounds of play. Suppose a sample of size N is drawn from S in order to play N rounds of the game. Conditional on the sample, E(SW) and E(PS) are finite and equal, so E(SW−PS) = 0, but this result is not spread uniformly throughout the sample. For instance, if the largest amount selected is 2^n, then switching on F_{n+1} means certain loss and at the highest stakes of the N rounds. More generally the negative expected value of switching on the few, typically sparse highest F_j's generated by the sample counterbalances the positive expected value of switching on the other F_j's.[4] Even though the finite expected values of SW and PS are equal, in a series of trials SW tends to generate positive expected value relative to PS for lower value and middle values of 2^j and negative expected value relative to PS for the highest values. A judicious switching strategy can exploit the bias.

Only the 4th approach accords with the fact that knowing the amount in the 1st envelope is useful for improving the player's results. Clark and Shackel judge opening the envelope and the information it reveals to be worthless. Meacham, Weisberg, Dietrich, and List ascribe a miraculous value to opening the envelope but find no use for the information it reveals.

We define two conditional strategies: $O(2^j)$ switches only on 2^j and otherwise passes; $M(2^j)$ switches on 2^j and all smaller amounts but passes on amounts greater than 2^j. $M(2^n)$ − PS has a positive expected value that increases with n (see Sect. 9.6). In Fig. 8.2 the dashed line represents E(M(32) − PS). It behaves like SW − PS up to and including S_5F_5, then behaves like PS − PS, thereby retaining the profitability of switching on S_5F_5. For large n, $M(2^n)$ − PS's expected gain of $(1−r)(2r)^n/2 \cong 0.125(1.5)^n$ depends on a windfall tied to a long shot. If n escapes to infinity, the windfall vanishes. In the limit SW − PS is equally balanced in SW's and PS's favor.

The two-envelopes game is not an unrealizable thought experiment—it can be played, painstakingly with envelopes, notes, etc., or rapidly through computerized simulation.[5] The latter requires simulation of drawing S_i from the geometric

---

(Footnote 3 continued)
should have known better declared the information revealed by the moderator to be useless for improving play.

[4] This is most easily seen in a large sample. Let $C(2^i)$ be the number of cases of $S = 2^i$ in the sample. Irrespective of the original probabilities, it is favorable over this particular sample to switch on 2^{i+1} if and only if $C(2^{i+1}) > C(2^i)/2$. This inequality tends to be true for the large counts that accompany small values of i, since $P_{i+1} > P_i/2$. But for the small counts that accompany the sparsely sampled highest values of i, it often happens that $C(2^{i+1}) < C(2^i)/2$, making it unfavorable to switch on 2^{i+1}.

[5] Meacham and Weisberg (2003) write as though expected value calculations and repeated trials are opposed in this problem: "it is misleading to speak of the expected utility (*EU*) of repeated trials, since in the peeking case the question is whether or not one should swap in a *particular* case, given that one has seen a particular amount in envelope *A*. ... one's decision regarding whether or not to swap in the peeking case should be determined by the EU of swapping for a particular value of *A* not on whether the EU of swapping is better 'on average' over repeated trials." This misses the possibility of using repeated trials to test not only SW but selective

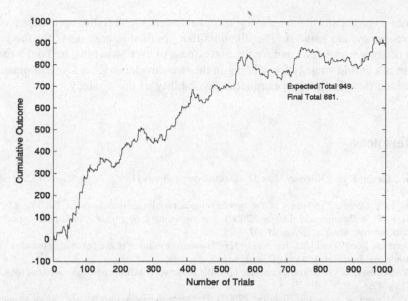

Fig. 8.3 This is a record of 1,000 trials of the strategy M(32) − PS, which made a total profit of 881; it represents the median outcome of 11 series, ranging in profit from 350 to 1485

distribution $P(S_i) = 0.25(0.75)^i$ and of tossing a coin to decide which envelope receives S_i and which $2S_i$. All that remains is player choice; however, we need to test strategies not people; and strategies are also easily automated. Despite the volume of theoretical disputation on the game, every matter in contention can be settled by means of experimental trials. I have tabulated millions of trials of this game. These confirm that SW is no better than PS and that for a large enough sample, a strategy such as M(32) − PS tends to perform near its expected value of $0.125(1.5)^5 \cong 0.949$ per trial. (see Fig. 8.3) The two-envelopes paradox is that rare philosophical problem that can be definitively resolved by means of computer.

In the two-envelopes game can be found a moral worthy of Aesop: reach for a giant portion and gain nothing; settle for a moderate share and succeed.

8.3 Summary

The case for switching envelopes under symmetrical conditions rests on equivocation. For the version in which one looks into the envelope before deciding whether to switch, new puzzles arise. Attempts to resolve this paradox can be categorized in terms of the truth values attributed to three key assertions. This

(Footnote 5 continued)
strategies that depend on the particular amount A. With an exact model as in the two-envelope problem, expected value calculations and repeated trials ought to *agree* to within sampling error.

creates a three part division among mistaken attempts. A fourth approach avoids inconsistency and paradox. The dilemma-like conflicting arguments for the benefits of always switching and for the uselessness of ever switching are both wrong. There is a hybrid strategy for winning in the two-envelopes game. An examination of eleven thousand trials confirms the favorability of this strategy.

References

Brams, Steven J. and Kilgour, Marc D., Discussion of Bruss (1996). *Math.Scientist, 23* (1998) 58–59.

Bruss, F. T. (1996). The fallacy of the two envelopes problem. *Math Scientist, 21,* 112–119.

Bruss, F. T., & Ruschendorf, Ludger. (2000). The switching problem and conditionally specified distributions. *Math Scientist, 25,* 47–53.

Christensen, Ronald and Utts, Jessica (1992). Bayesian resolution of the 'exchange paradox' *The American Statistician, 46* (4), 274–276.

Clark, Michael, & Schackel, Nicholas. (2000). The two-envelope paradox. *Mind, 109*(435), 415–442.

Clark, Michael, & Schackel, Nicholas. (2003). Decision theory, symmetry and causal structure: reply to Meacham and Weisberg. *Mind, 112*(448), 691–700.

Horgan, Terry (2000). The two-envelope paradox, nonstandard expected utility, and the intensionality of probability *No$\hat{u}$s* 34(4) 578–603.

Katz, Bernard D., & Olin, Doris. (2007). A tale of two envelopes. *Mind, 116*(464), 903–926.

Kretzmann, Norman, Anthony Kenny, and Jan Pinborg (eds.) (1982). The Cambridge history of later medieval philosophy, Cambridge University, Cambridge.

Meacham, Christopher J. G., & Weisberg, Jonathan. (2003). Clark and Shackel on the two-envelopes paradox. *Mind, 112*(448), 685–689.

Sobel, Jordan Howard. (1994). Two envelopes. *Theory and Decision, 36,* 69–96.

Chapter 9
Odds and Ends

9.1 Doomsday

Let A_0, A_1, A_2,... be a set of alternatives that are mutually exclusive and exhaustive, i.e., $P(A_i A_j) = 0$ if $i \neq j$, and $\sum_i P(A_i) = 1$. Let R be a condition, according to Bayes' Theorem.

$$P(A_i|R) = \frac{P(R|A_i)P(A_i)}{\sum_j P(R|A_j)P(A_j)}$$

This can be applied to the lottery example in Sect. 1.1. Let A_0 be the small lottery with ten numbers and A_1 the big lottery with a thousand numbers. Let R be the fact of drawing a seven. $P(A_0) = P(A_1) = 1/2$, $P(R|A_0) = 1/10$, $P(R|A_1) = 1/1,000$. By Bayes' Theorem,

$$P(A_0|R) = \frac{(1/10)(1/2)}{(1/10)(1/2) + (1/1,000)(1/2)} \cong 0.99$$

Drawing a seven shifts the likelihood it is the smaller lottery from 50 to 99 %.

Suppose A_i means "doom occurs when the total cumulative human populations reaches i" and R stands for the fact that one's birth rank is r. We then know $P(A_k) = 0$ for all k corresponding to people already born, since we know doom has not befallen us yet. HR requires a finite, albeit unknown number N, that is the total number of humans that ever live. If there exists no such number N, or if N turns out to be infinite, there can be no random selection from among N humans, and HR fails.

The crucial probability in this application of Bayes' Theorem is $P(R|A_i)$, the probability of having birth rank r *given* that doom will occur at cumulative population i. If, quite reasonably, one supposes that the probability of having birth rank r remains the same as one runs through various *later* doom scenarios, the Doomsday argument stops in its tracks; for then $P(A_i|R)$ is just $P(A_i)$ and the prior

W. Eckhardt, *Paradoxes in Probability Theory*, SpringerBriefs in Philosophy,
DOI: 10.1007/978-94-007-5140-8_9, © The Author(s) 2013

probabilities are left unrevised. To secure Doomsdayer conclusions, we need HR
from which it follows that $P(R|A_i) = i^{-1}$. Then

$$P(A_i|R) = \frac{i^{-1}P(A_i)}{\sum_j j^{-1}P(A_j)}.$$

Prior probabilities $\{P(A_i)\}$. shift to $\left\{ i^{-1}P(A_i) \Big/ \sum_j j^{-1}P(A_j) \right\}$ under the impact
of the HR assumption.

Suppose A_i and A_k are two doom scenarios with nonzero prior probability and
such that upon application of the transformation A_i gains in likelihood and A_k
loses; then

$$\frac{i^{-1}P(A_i)}{S} - P(A_i) > 0 > \frac{k^{-1}P(A_k)}{S} - P(A_k)$$

where $S = \sum_j j^{-1}P(A_j)$. Since $P(A_i)$ and $P(A_k)$ are positive $\frac{i^{-1}}{S} -$
$1 > 0 > \frac{k^{-1}}{S} - 1$. Then $1/iS > 1 > 1/kS$ and $i < k$. Therefore, it is always earlier
scenarios that gain likelihood from later scenarios.

Nielsen (1989, pp. 454–459] and Gott III (1993) have published similar arguments purporting to show that the time left until the end of the human race may be
shorter than we generally suppose. The treatments differ from Leslie's mainly in
emphasis. Gott's article has more mathematical trappings and brims with specu-
lation; he also considers questions other than that of human survival.

Leslie expends not a little effort countering the somewhat fatuous objection that
future humans are not alive to observe anything (Leslie 1996, pp. 19–21, 214–218,
246–247, 1993, pp. 489–490) in which, unaccountably, this objection is ascribed to
me. The issue is not what unborn humans can or cannot do, but what we can infer
about their numerosity from our birth rank.

Leslie's views on the regulating role of determinism in the Doomsday and
kindred arguments are unjustified. According to him, this kind of reasoning works
excellently under determinism, but as we slide along the scale of increasing
indeterminism Doomsday arguments become progressively undermined until for
the case of radical indeterminism they may fizzle altogether (Leslie 1992, pp. 537,
1996, pp. 188, 233, 234). However, if there existed a mode of statistical inference
that were valid according to the extent that determinism were true, then by
repeatedly testing the accuracy of this type of statistical inference, one could gauge
the correctness of determinism. Since this conclusion is highly implausible it is a
safe bet that statistical inferences, including those that underlie the Doomsday
argument, do not hinge on the truth of determinism. That is why the determinism
question is not a burning issue among say, insurance companies.

I would like to disentangle the problem from certain perennially unresolved
philosophical issues with which it has come to be associated. As long as the
validity of the Doomsday argument is made to hinge on whether the future is open

or fixed, or whether the future is fully implicit in the present, we can rest assured we are not going to settle the question of the argument's validity.

Leslie asserts that both the truth of indeterminism and its importance modulate the impact of the Doomsday argument. He claims that an open future "reduces the power of [doomsday] reasoning, instead of destroying it" (Leslie 1992, pp. 537) but gives as explanation only the possibility that indeterminism may not matter much to human survival. Yet all the key ingredients of the argument—Bayes' theorem, our birth rank, and our prior expectation of doomsday—are such that one cannot say why determinism should make a difference to them. Leslie does not mention determinism either in his central presentation of the argument or in numerous collateral examples. It is unclear what step of the argument a failure of determinism is supposed to weaken. In fact, the issue of determinism is a red herring. Determinist and indeterminist are on exactly the same footing when it comes to making probabilistic inferences. The practicing statistician need not be concerned with questions of whether physical process is ultimately deterministic or whether the future is open or fixed.[1]

9.2 The Betting Crowd

An event or a condition does not have a probability but rather numerous probabilities relative to numerous conditions. Conditional of being a member of the final population, the probability of wining is 1/10; conditional on being a member of a particular crowd, the probability of wining is 35/36. The question is which of these conditional probabilities is relevant to the player's prospects as she enters the crowd. It's the dice and not the final population that controls the player's outcome.

9.3 Sims

The Doomsday and Simulation arguments can be rephrased to make the retrocausal undercurrents explicit.

---

[1] The impression that determinism is relevant to the Doomsday argument may be motivated by the following inchoate reasoning: if I am to be a random member of the total human population, my expected rank needs to be the average human rank, but the average human rank depends on how many come after me. If the population to come after me were subsequently increased, say through the intervention of a benevolent angel, it would be unreasonable to suppose that my expected rank would be retroactively increased; we can only conclude that such unforeseeable additions to the human pool would compromise my status as random. The assumption of determinism serves to keep this potentially unruly future under control.

Doomsday: if we were not near the end of humanity's lifespan, the large implied future populations would have boosted our random birth ranks more than actual future populations did; therefore, we are near the end.

Simulation: if there ever are sims, their extraordinary population numbers would boost our random birth rank by such grandiose amounts that we ourselves would most likely be sims. Furthermore, if we are not sims, it is because there are insufficient sims in our future to boost random birth ranks. (This may shed light on Bostrom's claim "Unless we are now living in a simulation, our descendents will almost certainly never run an ancestor simulation" (Bostrom 2003, pp. 255).

9.4 Newcomb's Problem

The advisor sequence approach can only test problems in which there is room for disagreement about how to play. If all decision theories agree about the optimal choice, then one cannot obtain legitimate advisor or players that favor the inferior choice. This can happen in two ways. (1) The problem is causal, therefore all decision theories agree on expected values and hence on optimal play. In this case advisors can be replaced by a Bernoulli process B(r) that "advises" cooperation with probability r. In a causal problem the choice screens the outcome off from a player or advisor, so random manipulation of choice is sufficient. (2) The problem is not causal, therefore there is disagreement about expected values, but it is not great enough to produce disagreement as to optima. We cannot employ B(r) in this case because randomization of the choice variable can neutralize non-causal correlations. The solution is to add a constant—a fixed cost or reward to one of the choices so that in the revised problem how one treats noncausal correlations does tip the scales. (None of the problems we treat require this adjustment.) In this way advisor probabilities are assigned to every problem.

9.4.1 Alternatives to Advisors

Use of advisor sequences assures compliance with the advisory principle. Some alternatives to advisors fall short in this regard. In addressing the question of whether and to what extent outcome can be manipulated by choice, it would perhaps be more straightforward to randomize the choice directly, e.g., have the player flip a coin to make his decision. This procedure would serve to eliminate the common cause correlations in the Solomon story and the prisoner's dilemma, showing defection to be optimal. In Newcomb's problem there are two possible assumptions that can be made about the predictor:

1. The most reasonable is that the predictor cannot foresee the results of coin tosses in which case the coin toss would sever the correlation of player choice

to box contents. In a sufficiently long sequence of coin tossers, two-boxers would perform better than one-boxers. However, a player who two-boxed on the basis of this experiment would behave predictably and would most likely receive an empty 1st box. The experiment yields no coherent recommendation, e.g., take two-boxes, but toss a coin to decide this. In Solomon and the Prisoner's dilemma decision by coin flip violates no condition of the problem. In Newcomb's problem making the boxing decision by coin toss breaks the act-outcome correlation by neutralizing the paramount feature of a Newcombian game—the predictor's success rate. If the predictions are worthless, then of course two-boxing is optimal, but this is not Newcomb's problem. The befuddled player of the Solomon game or the prisoner's dilemma who resorts to tossing a coin, decides the original problem by this toss. The player in Newcomb's problem who resorts to a coin toss creates a new problem.

2. The predictor can foresee the results of coin tosses. In this case the choice-outcome correlations persist, forcing a one-boxing conclusion. But this is obtained through the artificiality of stipulating the coin tosses are predictable in the way human decisions are. Why not use human decision makers as advisors? Similar remarks, apply to decision theories, robots, artificial intelligences, extraterrestrials. They are either too predictable (decision theories, robots) or unpredictable in incomprehensible ways. Human advisors provide a better model than any of these.

9.4.2 The Consequence of Randomization

If randomization takes place at the level of advisor, through random sampling, this defines an advisor sequence. The result is the coherent theory. If randomization takes place at the level of player (for unadvised player sequences in which player preference matters) this defines a player sequence. The result is the evidential theory since player sequences leave intact all the act-outcome correlations that arise from the specifics of the problem. If randomization takes place at the level of the choice, say, by tossing a coin, the result is the causal theory. This is because the coin toss screens off any possible correlation owing to player or advisor such as common cause correlations; those that remain are purely causal. Finally randomization at the level of the outcome alters the game itself, since there remain no act-outcome correlations. In this case all decision theories make the same recommendation, e.g., if the contents of the 1st box is randomly chosen, all theories recommend two-boxing.

9.4.3 Liebnitzian Lethargy

Liebnitz famously maintained that monads (his term for ultimate individual entities) do not interact, rather each develops independently of the others. Apparent causal interactions among monads are maintained by pre-established harmony, i.e., the monads are synchronized to behave as though interacting. In our terminology all causal relations among separate monads are mimetic. Suppose a causal decision theorist Theo sits in an office in which there is a time bomb set to detonate. Since Theo and bomb are composed of different monads, the relationship of Theo to the bomb fits the definition of mimetic causation—Theo's behavior is indicative of the bomb's behavior, just as the player's choice is indicative of the box contents, but one does not cause the other. The bomb's exploding or not is a matter internal to its own monads which is already a fixed development irrespective of what Theo does. Hence Theo cannot justify the extra effort to dismantle the bomb. He would defect and get blown to bits. One can object that it is Liebnitzian causality that is bogus, not the dominance principle, but even so the example illustrates that where causality is simulated, causal decision theory is the wrong tool.

On a different topic dispute between evidential and causal decision theories has often been mischaracterized as a conflict between expected value and dominance reasoning. Each version of decision theory—evidential, causal, or the mediate theory outlined below, has its own version of expected value reasoning and each has its own dominance principle. The usual practice is to pit evidential expected value against causal dominance, so of course there is conflict. On the question of whether the choice of X or X + 1,000 is better, the **causal dominance principle** states that choosing X + 1,000 is better except possibly for cases in which the choice itself influences the value of X. The **evidential dominance principle** states that choosing X + 1,000 is better except possibly for cases in which the choice itself is probabilistically correlated to the value of X.[2] The corresponding expected value maximization principles are complimentary, not contradictory, to the dominance principles. The **causal (evidential) expected value maximization principle** states that the best choice is the one that maximizes expected value, taking into account all act-outcome correlations that are causal (evidential).

Proposition 1: If $R(T_1) \subset R(T_2)$, then T_2 has a narrower dominance principle and a broader expected value maximization principle than T_1.

---

[2] Causal dominance refers to a causal *exception* to dominance reasoning. Suppose I can steal a sum of money. If I don't go to jail, I'm better off with the money than without it; if I do go to jail, I'm better off with the money than without it. What makes this reasoning laughable is that it overlooks the causal relation between the theft and going to jail which precludes use of the causal dominance principle. Similarly the evidential dominance principle states that any kind of act-outcome correlation can provide exceptions to dominance reasoning.

Proof Relative to $R(T_1)$, $R(T_2)$ contains extra exceptions to the dominance principle and extra factors in expected value maximization. □

We endorse only the **coherent dominance principle** (exceptions are made for causal and mimetic influences) and the **coherent expected value maximization principle**.

9.4.4 Coherence Implies Stability

A problem that afflicts most decision theoretic formulations is **decision instability**: the decision calculation can vacillate without coming to a settled conclusion. Neither player nor advisor sequences can be unstable in this sense; they are both convergent sampling processes. We consider two examples (these are not cooperation problems). In the first (Gibbard and Harper 1978) the player has the choice of going to Damascus or Aleppo, in an attempt to avoid Death, but Death has predicted his choice and will be there to meet him. Vacillation results because when he considers choosing Damascus, Aleppo becomes the best choice and vice versa. The coherent solution is obvious: advisors who recommend Damascus would have players who fared just as poorly as those who were advised to go to Aleppo. Advisor sequences indicate this is a no-win situation with both alternatives equally bad. This is another illustration of the decision theoretic equivalence of causal and mimetic correlations. It should not matter to the player whether Death and the player correlate mimetically because Death predicts the player's choice or causally because the Death trails the player to the city.

The second is a game called "button" (Richter 1984). As in the prisoner's dilemma, this game has two players who are held incommunicado. The players are clones who are expected to act in the same way. Each must decide to push or not push a button. If both push, they each win 10. If one pushes and the other does not, they both win 100. If both refrain from pushing, they both lose 1,000. The instability results because as soon as the player decides to push and concludes his cohort will push, it becomes better not to push. Similar reasoning applied to the decision not to push, leads to the conclusion it is imperative to push, and back and forth it goes.

The evidential theorist should reason that the only outcomes are that both push or neither pushes, of which the first is much preferred, so the optimal choice is to push.[3] This argument overreaches as can be seen by introducing advisor

---

[3] Compare Davis's symmetry argument for the prisoner's dilemma (Davis 1985). Rational players necessarily play the same way. Of the four combinations that leaves only two: both cooperate or both defect. Of these the first is better, so cooperation is optimal. This argument overlooks that only *one* of the four combinations is rational, and this happens to be mutual defection.

sequences. Let P represent the designated player's decision to push, P_0 the other player's decision to push.

$$E(P) = P(P_0|P)10 + P(\overline{P}_0|P)100$$

$$E(\overline{P}) = P(P_0|\overline{P})100 - P(\overline{P}_0|P)1,000$$

Advisors eliminate clonal correlation. Let $r = P(P_0) = $ the fraction of the player population that pushes. The value of r can be estimated from advisor sequence data.

$$E(P) = 10\,r + 100(1-r)$$

$$E(\overline{P}) = 100r - 1,000(1-r)$$

Then $E(P) > E(\overline{P})$ if and only if $r < 0.924\ldots$ Pushing is usually optimal, but for $r > 0.924\ldots$ not pushing is best. Decision instability cannot occur in this format.

9.5 Is the Card Game at all Feasible?

Since there are earnest proposals for performing the Collider card game, we pause to address its feasibility. Many factors conspire to make this a futile exercise. A negative result would surely be meaningless—most likely it would not even prevent the experimenter from trying again. But what about a positive result? The parallels to Newcomb's problem were all drawn under the assumption that the radical theory as well as Nielson's interpretation of the mathematics were both correct. Without this assurance, justification of the card experiment presents formidable problems. There would be too many other ways to interpret a positive result (God, Satan, extraterrestrials, telekinesis, fraud) making it impossible to know if it were wise to cooperate.

In the standard Newcomb game the inducement to defect is relatively small; this is appropriate since the dominance principle is considered airtight by two-boxers, so a small inducement should be enough. A convinced one-boxer would presumably not find the strategy psychologically difficult to carry out. In OB it might be easy to intend to play the sport strategy and easy to recommend it sincerely, but it is decidedly more difficult to carry it out. In the Collider card game the temptation to defect is nearly irresistible. Assuming the Higgs anomaly is real, the only way CERN could run the card game so as to be likely to draw the jackpot card, would be to bind itself irrevocably to abide by its intention to close down the collider upon drawing the jackpot card. Expressed in causal terms the only way to obtain the million in OB, apart from predictor error, is to arrange matters so that reception of the million *causes* one-boxing. In the card game the only way to draw the jackpot card, apart from the negligible chance of accomplishing it randomly, is to arrange matters so that selection of the jackpot card *causes* abandonment of the collider.

9.6 Two Envelopes

$E(SW - PS|F_0) = 1$ and $P(F_0) = P_0/2 = (1 - r)/2$. For $i \geq 1$, F_i happens half the time S_i happens in which case $SW - PS$ looses 2^{i-1} and F_i happens half the time. S_{i-1} happens in which case $SW - PS$ makes 2^i. Hence $P(F_i) = P_{i-1}/2 + P_i/2 = (1-r)r^{i-1}/2 + (1-r)r^i/2 = r^{i-1}(1-r)(1+r)/2$

Cancelling the 2^{-1} factors,

$$E(SW - PS|F_i) = \frac{P_{i-1}2^i}{P_{i-1} + P_i} + \frac{P_i(-2^{i-1})}{P_{i-1} + P_i} = \frac{r}{1+r}2^i$$
$$- \frac{1}{1+r}2^{i-1} = \frac{2^i r - 2^{i-1}}{1+r}.$$

Since $M(2^n)$ passes on F_{n+1} and higher,

$$E(M(2^n) - PS) = \sum_{i=0}^{i=n} E(SW\text{-}PS|F_i)P(F_i)$$
$$= \frac{1-r}{2} + \sum_{i=1}^{i=n}\left(\frac{2^i r - 2^{i-1}}{1+r}\right)\left(\frac{r^{i-1}(1-r)(1+r)}{2}\right)$$
$$= \frac{1-r}{2} + \frac{(2r-1)(1-r)}{2}\sum_{i=1}^{i=n}(2r)^{i-1}$$
$$= \frac{1-r}{2} + \frac{-(1-2r)(1-r)}{2}\left(\frac{1-(2r)^n}{1-2r}\right)$$
$$= \frac{(2r)^n(1-r)}{2}$$

The value of $E(M(2^n) - PS)$ can also be derived as follows: for $i < n$, $M(2^n) - PS$ breaks even on all S_i. For $i > n$, $M(2^n) - PS = 0$. On S_n, $M(2^n) - PS$ makes 2^n with probability $P_n/2$ which equals $(2r)^n(1-r)/2$.

9.6.1 Additional Approaches?

Some authors pursue uniquely Bayesian analyses of the problem; these are distractions since the two-envelopes game has payoffs and probabilities that are fully specified and stipulated to apply exactly. This leaves no maneuvering room for obtaining different Bayesian and frequentist solutions, just as there is no disagreement as to the probability of heads for a perfectly unbiased coin.

Sobel (1994) and Blackman et al. (1996) make monetary boundedness a decisive factor in how to play. Real money is bounded, and even if we play using meaningless numbers, these are bounded by limitations on paper, ink, storage capacity, etc. It's not absurd to model the envelope amounts as unbounded; this is after all a problem in probability theory, not economics. What is absurd is to suppose an astronomical bound alters whether one should switch on a small amount. Should the analysis of a player whose 1st envelope contains 2^2 be driven by whether an envelope containing 2^{101} occurs once in two million years or instead never?

Meacham and Weisberg (2003) use A and B as symbols for the amounts in the 1st and 2nd envelopes respectively. The authors advocate **AL** admitting this "is bound to raise the old worry that if swapping is a good idea regardless of the value of A, it must be a good idea to swap even if one does not know the value of A. Thus one should swap in the no-peeking case." (p. 688). They seek to avoid this discomfiting conclusion by appeal to the difference between always switching in the informed game and always switching in the blind game. "inferring from the peeking case that swapping is a good idea in the no-peeking case amounts to... infer[ring] $E(B - A) > 0$ from the fact that $E(B - A|A= 2^n) > 0$ for all natural n." (p. 688). Weisberg and Meacham conclude "There is no inconsistency in maintaining that swapping is unhelpful.in the no-peeking case but beneficial in the peeking case." (p. 688).

Meacham and Weisberg claim that the strategy of always switching with knowledge of the amount in the 1st envelope is superior to the strategy of always switching without this knowledge. Readers for whom this is clearly ridiculous may wish to skip ahead to (9.6.2) to avoid the following *post mortem*.

In any play of the informed or blind game the outcome of switching depends on two numbers—the amount in the 1st envelope and the amount in the 2nd envelope. Merely opening the 1st envelope or learning its contents has no effect on either of these numbers; accordingly it can have no effect on the outcome of invariably switching. (Learning the contents of the first envelope can indeed be useful, but not to a player determined not to act upon this information.) In the idiom of Chap. 5, when the drawing from S and the envelope randomization are used as a shared probability source, the blind and informed games are in outcome alignment.

Let $\mathbf{E_1}$ be the statement "$E(B - A|A = 2^n) > 0$ for all n" and $\mathbf{E_2}$ the statement "$E(B - A) > 0$" Assume the envelopes in the blind game to be prepared the same way as in the informed game and that the player knows this. Let $\widehat{\mathbf{AL}}$ be the assertion concerning the blind game that **AL** makes concerning the informed game, namely that it is favorable to switch unconditionally. The authors' argument can be summarized as follows: **AL** raises the worry that $\mathbf{AL} \to \widehat{\mathbf{AL}}$, but $\mathbf{AL} \to \widehat{\mathbf{AL}}$ amounts to the fallacious $\mathbf{E_1} \to \mathbf{E_2}$; there is no inconsistency in $\mathbf{AL}\&\neg\widehat{\mathbf{AL}}$.

This reasoning relies on two misidentifications. First $\mathbf{E_1}$ is interpreted to mean **AL**. Since $\mathbf{E_1}$ is a direct statement of **EC** this amounts to tacit assumption of the fallacious $\mathbf{EC} \to \mathbf{AL}$. Second, $\mathbf{E_1} \to \mathbf{E_2}$ in the two envelopes game is taken to mean $\mathbf{AL} \to \widehat{\mathbf{AL}}$ instead of $\mathbf{EC} \to \mathbf{AL}$. This interpretation faces insurmountable difficulties: (1) $\mathbf{E_1} \to \mathbf{E_2}$ concerns the aggregation of individual switching strategies into one comprehensive switching strategy, whereas no aggregation occurs in

$\mathbf{AL} \to \widehat{\mathbf{AL}}$; both $\mathbf{AL}$ and $\widehat{\mathbf{AL}}$ refer to comprehensive switching strategies[4] (2) In the expression $\mathbf{E}_1 \to \mathbf{E}_2$ symbols such as A or B must be given the same meaning in each occurrence. In contrast $\mathbf{AL}$ and $\widehat{\mathbf{AL}}$ refer to distinct games. An accurate retort would be that the blind and informed games are sufficiently similar that the same symbols can be used in referring to each one. This leaves Meacham and Weisberg with a dilemma. If the symbols A and B can be consistently used in the informed and blind games, then the expression E(B − A) has the same value (or lack of value) for both games; therefore $\mathbf{AL}$ and $\widehat{\mathbf{AL}}$ have the same truth value, which is what the authors wish to deny. If instead the expression E(B − A) has different values in the informed and blind games, this belies the contention that the same symbols can be consistently used for both games. (3) Meacham and Weisberg appeal to the falseness of $\mathbf{E}_1 \to \mathbf{E}_2$, Dietrich and List (Dietrich and List 2005) to the falseness of a generalization of $\mathbf{E}_1 \to \mathbf{E}_2$ that they call the event-wise dominance principle,[5] and both articles refer to a more general formulation enunciated by Chalmers. These expressions are true in some cases and false in others, so as generalizations they are false, but this would not entail the falseness of $\mathbf{AL} \to \widehat{\mathbf{AL}}$ even if the latter were an instance of the former. In fact the three principles generalize the fallacious $\mathbf{EC} \to \mathbf{AL}$.

Dietrich and List (2005) defend two assertions: "Switch after opening together with the event-wise dominance principle contradicts indifference before opening" and "without the event-wise dominance principle there is no logical contradiction between switch after opening and indifference before opening." (p. 245) Switch after opening ($\mathbf{AL}$) contradicts indifference before opening ($\neg\widehat{\mathbf{AL}}$) for reasons such as everyday causal realism[6]; the dominance principle does not enter into it. Regarding the second assertion, if there is no logical contradiction between two policies or attitudes, then this surely remains the case after the removal of an irrelevant principle (the consistency of a formal system cannot be disrupted by removing an axiom). The second assertion may be technically correct, but the event-wise dominance principle is pure red herring.

This also give examples that purport to show that the perplexing $\mathbf{AL}\&\neg\widehat{\mathbf{AL}}$ is one of a class of similar results with the implication that this renders it less objectionable. They define two decision rules: of two options, **maximin** (often called minimax) takes the option with the smallest possible loss while **maximax** takes the option with the largest possible gain. If the two envelopes are prepared by independent drawings from the open interval (0, 1) then the maximiner is

---

[4] $\mathbf{E}_1 \to \mathbf{E}_2$ does imply $\mathbf{EC} \to \mathbf{AL}$. In the informed game, $\mathbf{EC} \to \mathbf{E}_1$ and $\mathbf{E}_2 \to \mathbf{AL}$, hence $\mathbf{E}_1 \to \mathbf{E}_2$ implies $\mathbf{EC} \to \mathbf{AL}$. In going from $\mathbf{EC}$ to $\mathbf{AL}$ individual cases are aggregated into one case, just as in going from $\mathbf{E}_1$ to $\mathbf{E}_2$.

[5] "Let P be a partition of the setoff all possible states of the world into non-empty events. For any two lotteries L_1 and L_2 conditional on observing event E for every E in P, then you strictly prefer L_1 *to* L_2 unconditionally".

[6] All that needs to be ruled out is magic with no magician and no explanation, physical or extra-physical. Prehistoric animists would probably consent to this exclusion.

indifferent to switching before, but always against switching after he opens his envelope. The maximaxer is indifferent to switching before, but always in favor of switching afterwards.

In spite of its importance in two-person game theory, maximin can be made to give foolish answers in decision problems in which there is no opponent and consequently no good reason to concentrate exclusively on the worst case (Cox and Hinckley, p. 434). Maximin and maximax are both known to give arbitrarily bad advice in certain contrived decision problems, e.g., the maximiner will pass up a good chance at a million dollars to shave a penny off his possible loss; the maximaxer will risk a million dollars to add a penny to his possible gain (for other examples see (Berger 1980, pp. 371–376). We focus on the maximaxer since it most closely fits the analogy with $\mathbf{AL}$ and $\widehat{\mathbf{AL}}$ that the authors wish to press. If the maximaxer finds 0.9999 in his envelope, the maximax rule recommends switching since the unknown envelope has a higher possible profit. Yet in switching he has a very small chance to gain a miniscule amount and a large chance to lose hundreds or thousands of times more. The maximaxer does not hold that always switching is *better* in the sense that it gives better outcomes. He follows a rule that recommends switching even though it is not better. These are excellent examples of the failure of maximin and related decision rules; like other such examples they do not transfer to expected value or expected utility maximization.

9.6.2 Causal Structure

In their attack on $\mathbf{EC}$ Clark and Shackel intimate that $\mathbf{EC}$ results from a misstep similar to the ones causal decision theorists see in evidential treatments of Newcomb's problem or the prisoner's dilemma. Causal decision theory however does not routinely disagree with the evidential varieties. If they were in disagreement about, say, whether it's profitable to play casino Roulette on an unbiased wheel, we wouldn't need esoterica such as Newcomb's problem or the Solomon story to decide between them. In the cases that have conflicting evidential and causal solutions, such as the Prisoner's Dilemma or Newcomb's problem, the evidential reasoner uses his own choice as evidence of what the other prisoner will do or what the predictor has done. This is without question an incorrect procedure in the Prisoner's Dilemma. In Newcomb's problem, the waters are muddied by the activity of the mysterious and confounding Newcombian predictor. There is nothing remotely like this in the proof of $\mathbf{EC}$, which depends on routine expected value calculations like those for Roulette. Despite the aura of paradox, decisions in the two-envelopes game belong to the majority for which causal and evidential theories concur.

Clark and Schackel contend that the "correct" series is the one that corresponds to the causal structure of the game that this is uniquely the S-series. Useful decision theoretic formulas should observe causal aspects of the game, but this does not mean that all relevant causal aspects can be encapsulated in a single series. The possibility and profitability of switching only on F_j also deserves to be considered part of the causal structure of the game. This is recorded in the F-series.

9.6.3 The Finite Two-Envelopes Game

Virtually all participants in the two envelopes debates agree on one matter: the paradox turns on the infinitary features of the game, especially the infinite expected values and the attendant divergences. The standard infinitary version may be more striking or more elegant, but relevant features of the informed game paradox can be presented in a finitary setting in which it is easier to uncover non sequitors and reconcile conflicting strands of the argument. The correct approach can be studied "in miniature" and transferred to the infinite case. For contrast we refer to the informed game considered up to this point as the **infinite game**. The **finite game** results through stipulating a maximum possible amount for S: all amounts from 2^0 to 2^{99} occur with the same probability as before $\left(P(S_i) = P_i = 0.5(0.75)^i, i < 100\right)$ while 2^{100} receives the remaining tail probability[7] of $4P_{100}$ (for $i > 100$, $P(S_i) = 0$; $P(S_{100}) = 4P_{100} = (0.75)^{100} \cong 3.2 \times 10^{-13}$). This makes $F = 2^{101}$ so rare it would take on average about two million years for it to occur if the game were played once every second.

EC says every $O(2^j)$ is favorable. This is true in the finite game *except* for $O(2^{101})$, the preposterous policy of switching *only* on 2^{101} (which as a practical matter is indistinguishable from PS). So **EC** is true in each case except for one of absurdly low probability.

SYM holds exactly as in the infinite game. Symmetrical ignorance applies irrespective of how the randomized envelopes are prepared.

AL says SW is better than PS. In the infinite game the divergence of $E(SW - PS)$ makes it harder to state the sense in which SW and PS are equivalent and perhaps easier to evade the consequences of this equivalence. In the finite game **AL** is false for elementary reasons. $E(SW) = E(PS) = \sum_{i=0}^{i=n} P(S_i)(2^i + 2^{i+1})/2$ a finite sum. SW and PS are equally good. In the finite game **EC** is nearly universally true while **AL** is outright false, a good finite approximation to the falseness of **EC** $\rightarrow$ **AL** in the infinite game.

Since $E(SW - PS)$ forms a convergent series, grouping the terms does not change the sum. Grouping terms with the same value of S yields a series of zeroes, summing to zero, and grouping terms with the same value of F, yields a series with increasing positive terms, summing higher and higher, except there is one last

---

[7] The probability of exceeding a certain value, called a tail probability, takes a simple form in the geometric distribution: $P(S > 2^n) = r^n$ (Balakrishnan and Nevzorov, p. 64) for $r = 0.75$, this is $(0.75)^n = 4P_n$.

negative term, corresponding to switching on $F = 2^{101}$, that cancels all previous terms and brings the sum to zero. The give-back on $F = 2^{101}$ makes the invalidity of **AL** transparent and the paradox evaporates. In the infinite game, the give-back is repeated endlessly but never definitively.

Every pertinent feature of the informed game paradox and of its resolution survives the transition to the finite game. On any reasonable time scale the outcomes of finite and infinite games are identical with virtual certainty. (For instance in the eleven thousand trials mentioned above, the highest value of S selected was 2^{37}. The negligible difference between playing the finite and infinite games would not have affected these results.) A flukish exception at the fringes of possibility may enter into a rigorous calculation but should not affect practical play. The infinite game displays the paradox in a starker exception free form, but the finite game reveals its structure in a context that avoids the subtleties and complications of infinite series.

9.6.4 Ross's Theorem

Ross (1994) proved a result that's played an interesting role in interpretations of this paradox. The theorem, not itself in doubt, has been misinterpreted in ways much like the paradox from which it sprang. We've seen that some forms of $M(2^n)$ are good strategies in the informed game but this depends crucially on the fact that $P_i < 2P_{i+1}$ for all i. Brams and Kilgour (Brams and Kilgour 1998) claimed this rule was "shown to be optimal" Blackman, Christensen, and Utts (Blachman et al. 1996) that "even without a prior distribution" one "can do better than always trading envelopes". Although undoubtedly of interest in its own right, the theorem gives no guidance whatever as to how to play a two-envelopes game. Ross's Theorem does demonstrate that knowledge of the amount in the 1st envelope can be exploited for profit, but this can already be seen in the ordinary version of the game. The theorem is not informative about how to play; in particular it does not imply that M(t) is in any sense a "good" strategy.

Ross proves the theorem for probabilistic (or mixed) strategies and for continuous money. The issue however can be well encapsulated using the discrete monetary amounts 2^i and the deterministic strategy $M(2^n)$. Let $\{p_i\}$ be *any* probability distribution on $\{2^i\}$, $\sum_{i=0}^{\infty} p_i = 1$. 2^i and 2^{i+1} are placed in the envelopes with probability p_i.

Ross's Theorem (discrete version). If $p_i \neq 0$, $E(M(2^i) - ST) > 0$.

Proof If 2^j is drawn with j < i, then $M(2^i) - ST$ breaks even on average since the strategy switches on 2^j and 2^{j+1}. If 2^k is drawn with k > i, $M(2^i) - ST$ is exactly zero. If 2^i is drawn, $M(2^i) - ST$ switches on 2^i and passes on 2^{i+1} for a gain of 2^i. $E(M(2^i) - ST)$ is then $2^i p_i/2 = 2^{i-1} p_i$. $\square$

A good example is $p_i = (0.75)(0.25)^i$, a geometric distribution with r = 0.25.

The formulas from (9.6) apply, so for $i > 0$ $E(SW - PS|F_i) = (2^i(0.25) - 2^{i-1})/(1.25) = (2^{i-2} - 2^{i-1})/(1.25) = -2^{i-2}/1.25 < 0$. Yet $E(M(2^i) - PS) = (0.5)^i(0.75)/2 = 0.375(0.5^{-i}) > 0$ in accordance with Ross's Theorem. Although switching is unfavorable in every case except $F = 2^0$, $M(2^i)$ is always better than PS. (It should be noted how paltry this expected value can be. $E(M(2^{10}) - PS) \cong 0.00037$, even though in the crucial cases the envelope contains 1024 or 2048.) Switching on the lowest possible amount is always favorable; however, there need not be a lowest possible amount, e.g., there could be arbitrarily small positive amounts but no zero amount, or there could be arbitrarily low negative amounts. One can therefore find distributions for which switching is unfavorable in every single case, yet $M(x) - PS$ makes a profit for every possible amount x. This can be considered a third Two-Envelopes paradox.

References

Berger, J.O. (1980). *Statistical Decision Theory and Bayesian Analysis.* New York: Springer.

Blachman, N.M., Christensen, R., & Utts, J.M. (1996). Comment on Christensen and Utts (1992). *The American Statistician 50*(1), 98–99.

Bostrom, N. (2003). Are you living in a computer simulation? *Philosophical Quarterly, 53*(211), 243–255.

Brams, S.J. & Kilgour, M.D. (1998). Discussion of Bruss (1996). *Math Scientist 23,* 58–59.

Cox, D.R. & Hinckley, D.V. (1974). *Theoretical statistics,* Chapman and Hall, London.

Davis, Lawrence H. (1985). "Is the symmetry argument valid?", in Campbell, R., and Sowden, L. (1985). *Paradoxes of rationality and cooperation,* University of British Columbia, Vancouver; 1985, 255–263.

Dietrich, F. & List, C. (2005). The two-envelope paradox: An axiomatic approach *Mind, 114,* 454, 239–248.

Gibbard, A. & Harper, W. (1978). Counterfactuals and two kinds of expected utility. *Foundations and Applications of Decision Theory,* Hooker, Leach & McClennen (eds.) vol. 1, D. Reidel, Dordrecht, 125–162. Reprinted in Gardenfors & Sahlin; 1988, and with abridgement in Cambell & Sowden; 1985.

Gott, J. R. (1993). Implications of the copernican principle for our future prospects. *Nature, 363,* 315–319.

Leslie, J. (1992) Time and the anthropic principle. *Mind,* 101, 403, 521–540.

Leslie, J. (1993). Doom and probabilities. *Mind 102,* 407, 489–491.

Leslie, J. (1996). *The End of the World.* New York: Routledge.

Meacham, C. J. G., & Weisberg, J. (2003). Clark and Shackel on the two-envelopes paradox. *Mind, 112*(448), 685–689.

Nielsen, H.B. (1989) Random dynamics and relations between the number of fermion generations and the fine structure constants. *Acta Physica Polonica, B20*(5), 427–468.

Richter, R. (1984). Rationality revisited. *Australasian Journal of Philosophy, 62,* 392–403.

Ross, S. M. (1994). Comment on Christensen and Utts (1992). *The American Statistician, 48*(3), 267.

Sobel, J. H. (1994). Two envelopes. *Theory and Decision, 36,* 69–96.

Epilogue
Anthropic Eden

We consider three stories Bostrom (2002, pp. 142–150) calls the **Adam and Eve Experiments**, concerning a couple who for unspecified reasons are the only two people to exist so far. Bostrom refers to such stories as "paradoxes of the self-sampling assumption" but I see them as refutations of Anthropic reasoning.

1st experiment: if Adam and Eve reproduce, they will be expelled from the garden and subsequently have billions of descendants. A Serpent advises them that they can mate with abandon since any reproduction would result in their having such a preposterously low birth rank among the billions that follow that successful reproduction is nearly ruled out.

2nd experiment: this time they form the firm intention to have a child unless a wounded deer limps into their cave. Bostrom equips the cave with advanced in vitro fertilization technology, evidently to assure there are no slip-ups in this regard. Their birth rank if they reproduce will be so improbably low that it's reasonable to expect the deer to limp in.

3rd experiment: the couple decide to retro-actively make the top card in a deck shuffled this morning be the queen of spades, using the same threat as above.

For simplicity it is assumed Adam and Eve either do not reproduce at all or have billions of descendants. In the first case with only two possible birth ranks, they possess likely ranks; in the second case their birth ranks are wildly improbable. The prevention of this improbability is said to bring about the sterility, prompt the entrance of the limping deer, or permit retroactive stacking of the deck. This bears a striking resemblance to the Higgs anomaly with babies in place of bosons, except that Nielsen and Ninomiya provide a physical theory to account for retrocausal effects, Bostrom attempts this through HR gymnastics alone. (After intricate peregrinations Bostrom concludes a portion of the absurdity can be averted without abandoning HR).

True to tradition, the Serpent lies. It relies on that old devil—human randomness—to persuade the naive couple. In their unique circumstances it is especially fatuous for Adam and Eve to be considered random individuals. According to these stories there will be large future populations only if Adam and Eve reproduce, but according to HR large future populations render such reproduction highly unlikely. Large future populations make their own cause unlikely. This snarled conclusion indicates that once again human randomness reasoning has gone awry. The absurdities of limping deer and card trick are also illustrative nonsequitors. Let R be reproduction, and D the limping deer (or the card trick). HR implies P(R) is near zero. Adam and Eve intend to keep P(R or D) very near one. The only outlet then is for P(D) to be near one. However, they can keep P(R or D) near one only by making R very likely if D doesn't happen. If D doesn't happen, the HR-induced low probability of R makes it impossible to keep P(R or D) near one. The high-tech equipment doesn't help: the easier is for them to reproduce, the greater the conflict with HR.

All three stories betray the Original Sin of Anthropic reasoning: using probability concepts in disregard of their causal implications. In any event, having begun with Doomsday, it's fitting to end with the Garden of Eden.

Reference

Bostrom, N. (2002). *Anthropic Bias; Observation Selection Effects in Science and Philosophy*. Routledge, New York.

Index

A

Adam and Eve experiments, 75–76
Advisors, 26–28, 31–32, 36,
    62–63, 65–66
Advisory principle, xiv, 27, 62
Advisor probabilities, 31, 33, 36, 62
Advisor sequence, 27, 31–33, 62–63,
    65–66
AL, 50–55, 68–71
Anthropic Eden, 75–76
Anthropic reasoning, xiv, 16, 75–76

B

Bayes' Theorem, 1, 8, 59, 61
Bayesianism, xiv, 8
Bernoulli process, 28–30, 33, 62
Betting crowd, xiii, xiv, 2, 11–13, 16–17,
    21, 61
Birth rank, 7–10, 15–17, 59–62, 75
Black box data, 33–34
Blachman, N. M., 72–73
Bostrom, Nick, 7, 10, 15, 17, 62, 73, 75–76
Brains in robots, 15–16
Brams, Steven J, 52, 58, 72–73
Bruss, F. Thomas, 48, 55, 58, 73

C

Carter, Brandon, 1
Causal problem, 28, 47, 62
Chalmers, David, 69
Christensen, Ronald, 55, 58, 72–73

Clark, Michael, 48, 52–56, 58, 70, 73
Coherence principle, xv, 30–34, 65
Conditional independence, 25
Cooperation problem, 30–32, 65
Correlation, 10, 22–23, 25–26, 28–29,
    31–33, 50, 62–66
    causal, 25–26 , 33
    chance, 26
    common cause, 25–26 , 62–63
    mimetic, 33, 40, 42–43, 50, 64–65
    positive, 10
    specious, 33

D

Davis, Lawrence H, 65, 73
Decision instability, 65–66
Decision theories, 21, 23–24, 28, 34, 43,
    55, 62–64
    acceptable, 26–27
    causal, 23–24 , 31, 34, 41, 64, 70
    coherent, 24, 32–33
    evidential, 23–24
    reformed evidential, 24, 31
Defection, 22–23, 30, 32, 35–37, 40–41, 43,
    47, 62, 65
Dependence, 8, 10–11
Deterministic problem, 28–29
Dietrich, Franz, 53, 56, 69, 73
Dominance principles, xiv, 47, 64
    causal, 64
    coherent, 65
    evidential, 64

W. Eckhardt, *Paradoxes in Probability Theory*, SpringerBriefs in Philosophy,
DOI: 10.1007/978-94-007-5140-8, © The Author(s) 2013

Dominance reasoning, 38, 64
Doomsday paradox, xiv, 1, 2, 7–11, 13,
        15–17, 59–62, 76

E
EC, 50–55, 68–71
Expected value
        maximization, 64–65, 70
    causal, 64 , 65
    coherent, 65
    evidential, 64
Extended randomness
        assumption, 15
External probability source, 28–29

F
Finite two-envelopes game, 71
Frequentism, xiv

G
Gott, Richard J., 60, 73

H
Hadron collider card
        experiment, 3, 39–44
Higgs anomaly, 40–41, 43, 66, 75
Higgs bosons, 3, 39–44
Horgan, Terry, 55, 58
Human randomness assumption, 8, 9, 47,
        75–76

J
Jackpot card, 39–43, 66
Jeffery, Richard, 24, 28, 34

K
Katz, Bernard D, 48, 55, 58
Kilgour, Marc D, 52, 58, 72–73

L
Leslie, John, 1, 2, 4, 8–10, 60–61, 73
List, Christian, 53, 56, 69, 73

M
Maximax, 69–70
Maximin, 69–70
Meacham, Christopher, 53, 56, 58,
        68–69, 73

N
Newcomb's problem, xiv–xv, 2–3, 21–35,
        41–43, 50, 62–66, 70
Nielsen, Holger B, 39, 44, 60, 73, 75
Non-causal problem, 28–29, 32–33, 62

O
Odysseus and the Sirens, 36–37
Olin, Doris, 48, 55, 58
One-boxer, 2–3, 21–24, 28–32, 35–37,
        42–43, 63, 66
Open box game, xiii–xiv, 3, 35–38, 40–43
Outcome alignment, xv, 28–31, 33, 68

P
Player probabilities, 31–32, 36
Player sequence, 27, 31–32, 63
Prisoner's dilemma, 23–24, 26, 30, 34,
        62–63, 65, 70
Probabilistic independence, 25
Probabilistic problem, 29

R
Randomness in a reference
        class, 8–9, 12
Retrocausality, 8–9, 13, 16, 17
Ross's theorem, 72–73
Ruschendorf, Ludger, 48, 55, 58

S
Screening off, 25
Shackel, Nicholas, 48, 53, 56, 58, 70, 73
Sims, 2, 15–17, 61–62
Simulation Argument, 2, 7, 11, 13,
        15–17, 61
Sobel, Jordan Howard, 52, 58, 67, 73
Solomon story, 23–24, 26, 30, 32, 43,
        62–63, 70

SYM, 50–58
Symmetrical ignorance
    principle, xiv, 48, 50–51, 53

T
Tribal PD, 23, 32, 43
Two-boxer, 3, 22–24, 28–31, 35–37,
    43, 63, 66
Two envelopes paradox, xiii–xv , 3–4,
    47–58, 67–73
  the blind game, 3–4, 47, 50, 51, 53, 55,
    68, 69

the informed game, 4, 49–50, 53, 55,
    68–69, 71–72

U
Utts, Jessica, 55, 58, 72–73

W
Weisberg, Jonathan, 53, 56,
    58, 68–69, 73